avian

reservoirs

EXPERIMENTAL FUTURES

Technological Lives, Scientific Arts, Anthropological Voices
A series edited by Michael M. J. Fischer and Joseph Dumit

avian reservoirs

VIRUS HUNTERS & BIRDWATCHERS IN CHINESE SENTINEL POSTS

FRÉDÉRIC KECK

DUKE UNIVERSITY PRESS
Durham and London 2020

Printed in the United States of America on acid-free paper ∞
Designed by Aimee C. Harrison
Typeset in Adobe Caslon Pro by Copperline Books

Cataloging-in-Publication Data is available from the
Library of Congress.
ISBN 978-1-4780-0613-8 (hardcover : alk. paper)
ISBN 978-1-4780-0698-5 (pbk. : alk. paper)
ISBN 978-1-4780-0755-5 (ebook)

Cover art: Composite of photo of Hong Kong skyline, by David Iliff, accessed at Wikimedia Commons; drawing from *Anatomy, Physiology, and Hygiene*, by Jerome Walker, accessed at Wikimedia Commons; electron microscope scan "Negative Transmission," by Cynthia Goldsmith and Dr. Terrence Tumpey, USCDCP, accessed at pixnio.com; photo of egrets in flight at Pocharam Lake, by J. M. Garg, accessed at Wikimedia Commons.

CONTENTS

acknowledgments vii
introduction 1

part I
ANIMAL DISEASES

CHAPTER ONE culling, vaccinating, and monitoring contagious animals 11

CHAPTER TWO biosecurity concerns and the surveillance of zoonoses 29

CHAPTER THREE global health and the ecologies of conservation 44

part II
TECHNIQUES OF PREPAREDNESS

CHAPTER FOUR sentinels and early warning signals 69

CHAPTER FIVE simulations and reverse scenarios 108

CHAPTER SIX stockpiling and storage 139

conclusion 173 – notes 179 – bibliography 211 – index 237

ACKNOWLEDGMENTS

As viruses mutate silently in animal reservoirs before emerging among humans, this book is the product of a long maturation among several host institutions and a few public outbreaks.

My work on pandemic preparedness has been initiated by the research project launched by Stephen Collier, Andrew Lakoff, and Paul Rabinow on biosecurity. I have been illuminated by discussions with Carlo Caduff, Lyle Fearnley, Stephen Hinchliffe, and Limor Samimian-Darash on the implications of avian influenza in this emerging field. The workshop we organized with Christopher Kelty and Andrew Lakoff on sentinel devices was a catalyzing moment in my reflection.

This book is based on long-term fieldwork at the Pasteur Centre of the University of Hong Kong. I want to thank its two directors, Roberto Bruzzone and Malik Peiris, for their welcoming and stimulating discussions. Isabelle Dutry, Jean-Michel Garcia, Martial Jaume, Nadège Lagarde, Jean Millet, Béatrice Nal, Dongjiang Tang, and Huiling Yen have patiently answered my questions and shared their practices. François Kien has designed the image for the conference I organized in 2009, "Avian Flu: Social and

Anthropological Perspectives," which has been an inspiration for the cover of this book. Robert Peckham has been a constant interlocutor through the Centre for the History of Medicine he has built at the University of Hong Kong. Gavin Smith has allowed me to build bridges between Hong Kong and Singapore, where he set up a Programme in Emerging Infectious Diseases at Duke-NUS Medical School. I benefited from the network of social scientists connected by the French Center for Research on Contemporary China, and I want to thank Jean-François Huchet and Paul Jobin, who directed its Hong Kong and Taipei branches when I visited them. The Hong Kong Birdwatching Society has encouraged me to compare their activities with those of birdwatchers in Taiwan, particularly through the stimulation of Mike Kilburn. I thank Mary Chow at the Agriculture, Fisheries and Conservation Department of the Hong Kong Government for her interviews and for the authorization to reproduce the map of a poultry farm.

The Laboratory for Social Anthropology at the Collège de France has hosted me for the last ten years, and allowed me to assume my intellectual debt to Claude Lévi-Strauss. Philippe Descola has constantly supported my shift from philosophy to sociology and anthropology, and I have benefited from the renewal of the structuralist program by colleagues such as Laurent Berger, Julien Bonhomme, Pierre Déléage, Andrea-Luz Gutierrez-Choquevilca, Perig Pitrou, and Charles Stépanoff. I thank Carole Ferret for exploring with me the field of animal studies in the seminar we have organized together for ten years. My research on zoonoses and human/animal relations at the Laboratory for Social Anthropology has been supported by grants from the Fyssen Foundation and from the Axa Research Fund. It led to creative collaborations with Nicolas Fortané, Vanessa Manceron, Arnaud Morvan, Sandrine Ruhlman, and Noelie Vialles. Christos Lynteris has been a strong and reliable partner in building a social anthropology of zoonoses, as well as Hannah Brown, Ann Kelly, and Alex Nading.

The project "Antigone," led by Thijs Kuiken at the University of Rotterdam in 2012–16, has opened a window on the European practices of biosecurity and pandemic preparedness. Marion Koopmans has helped me build bridges between virologists, epidemiologists, and anthropologists around zoonotic outbreaks, such as the emergence of MERS-CoV among camels in Qatar that we covered with Sarah Cabalion.

The project on simulations of disasters at Sciences Po in 2011–12 was a flexible framework for fruitful collaboration with Sandrine Revet and Marc Elie. The fictional and ritual aspects of the perceptions of disasters were at the heart of discussions we had with Mara Benadusi, Guillaume Lachenal, Katiana le Mentec, and Vinh-Kim Nguyen.

Between 2014 and 2018, my position as the head of the research department of the musée du quai Branly has oriented my reflections toward the classification and conservation of collections. I want to thank Stéphane Martin and Anne-Christine Taylor for trusting me in that unexpected position, and those who made possible the life of our department amid a range of scientific and cultural activities: Julien Clément, Marine Degli, Jessica de Largy Healy, Anna Laban, Maïra Muchnik, and Erika Trowe. The seminar we have organized with Tiziana Beltrame and Yaël Kreplak on the ecologies of collections has been inspirational for the writing of chapter 3 in this book.

In the same period, I have become a fellow of the Canadian Institute for Advanced Research in the program Humans and the Microbiome, which allowed me to explore new kinds of relations between humans, animals, and microbes. This program has strengthened older collaborations with Tamara Giles-Vernick and Tobias Rees, and fostered new ones with Brett Finlay, Philippe Sansonetti, and Melissa Melby.

The manuscript of this book was presented for my Habilitation to Research Direction in April 2017 in front of a jury composed by Philippe Descola, Didier Fassin, Jean-Paul Gaudillière, Yves Goudineau, Sophie Houdart, Annemarie Mol, and Anne-Marie Moulin. I want to thank them again for the intellectual exchanges we have had at the crossroads of history of sciences, philosophy of medicine, and anthropology of biopolitics.

I was fortunate to be invited to present my work in academic seminars or conferences by colleagues and friends who shared their comments and remarks: Shin Abiko, Warwick Anderson, John Borneman, Tanja Bogusz, Thomas Cousins, Ludovic Coupaye, Hansjorg Dilger, Paul Dumouchel, Hisashi Fujita, Dan Hicks, Cai Hua, Emma Kowal, Eduardo Kohn, Hannah Landecker, Javier Lezaun, Wang Liping, Nicholas Langlitz, Rebecca Marsland, Laurence Monnais, Anand Pandian, Joanna Radin, Hugh Raffles, Joel Robbins, Miriam Ticktin, Stefania Pandolfo, Anna Tsing, Meike Wolf, Kozo Watanabé, Jerome Whitington, Tang Yun, Patrick Zylberman. I also want to thank Luc Boltanski, Vincent Debaene, Emmanuel Didier, Nicolas Dodier, Marie Gaille, Isabelle Kalinowski, Patrice Maniglier, and Frédéric Worms for the discussions we have had over the different turns of my intellectual curiosity.

At Duke University Press, Kenneth Wissoker and Michael Fischer have strongly supported my French perspective in the debate on preparedness.

I am grateful to Anitra Grisales who edited my first manuscript, and to Susan Albury, Nina Foster, Aimee Harrison, and Colleen Sharp who followed the next steps of the editorial process.

My wife Joelle Soler has accompanied me in my travels and thinking, and helped me to keep my orientation. Our children Sylvia and Rafael have made our home a place of curiosity and wonder.

Prior versions of some of the material for this book appeared in the following articles: "A Genealogy of Animal Diseases and Social Anthropology (1870–2000)," *Medical Anthropology Quarterly* 33, no. 1 (2018): 24–41 (chapter 1); "L'alarme d'Antigone: Les chimères des chasseurs de virus," *Terrain* 64 (2015): 50–67 (chapter 2); "Avian Preparedness: Simulations of Bird Diseases in Reverse Scenarios of Extinction in Hong Kong, Taiwan and Singapore," *The Journal of the Royal Anthropological Institute (incorporating Man)* 24, no. 2 (2018): 330–47 (chapter 5); "Stockpiling as a Technique of Preparedness: Conserving the Past for an Unpredictible Future," in *Cryopolitics: Frozen Life in a Melting World*, ed. Joanna Radin and Emma Kowal, 117–41 (Cambridge, MA: MIT Press, 2017) (chapter 6).

INTRODUCTION

An influenza pandemic is one of the events that raise concerns at the global level. The cyclical character of flu pandemics—the 1918 "Spanish flu," the 1957 "Asian flu," the 1968 "Hong Kong flu"—leads experts to think that a new pandemic is imminent and that it would kill millions of people.[1] The question, according to global health authorities, is not when and where the pandemic will start, but if we are prepared for its catastrophic consequences. Pandemics disrupt social life not only because they kill individuals but also because contagion triggers panic and mistrust. Hence the need to be prepared for pandemics to mitigate not only their human casualties but also their social aspects.

Pandemics start when new pathogens infect a nonimmunized human population. It is considered that microbes mutate across animal species, developing usually without symptoms in their "animal reservoirs" before jumping to humans, in which they cause infection and contagion. Influenza viruses mutate and reassemble among birds, particularly waterfowl considered as "sane carriers" because they transmit the virus without being infected by them, and pigs, described as "mixing vessels" because they have receptors in their respiratory tracts that bind to bird viruses and human viruses. When microbiologists follow pathogens in their animal reservoirs to anticipate their emergence among humans and understand how they shift from "low pathogenic" to "high pathogenic," they introduce animals into the social.

This book asks, with the methods of social anthropology, how techniques to prepare for influenza pandemics have transformed our relations to birds. Billions of poultry have been killed all over the world to eradicate potentially pandemic pathogens from jumping over the species barriers. Migratory birds have been monitored to understand the spread of flu viruses outside of their place of emergence. Wild waterfowl have moved from the nature pages of magazines and newspapers to the front pages of major news coverage, depicting bird flu outbreaks as if they were terrorist attacks, while images of chickens in slaughterhouses have invaded the public space to ambivalently reassure consumers that chicken meat is safe to eat.[2] If the deadly pandemic bird flu virus still remains to come, its anticipation has already modified the world in which humans live with animals, wild and domestic.

Bird flu is described as a "zoonosis," an infection caused by a pathogen that jumped from nonhuman animals to humans. The concern for zoonoses, which constitute the main part of emerging infectious diseases, has grown in the last forty years with the fight against Ebola hemorrhagic fever (1976), transmitted from bats to monkeys, mad cow disease (or Bovine Spongiform Encephalopathy, 1996), transmitted from sheep to cows, and SARS (Severe Acute Respiratory Syndrome, 2003), transmitted from bats to civet cats.[3] While the link between pathogens and their environment has always been at the heart of public health, this series of emergences in the last forty years was explained by the dramatic changes in urbanization, deforestation, industrial breeding, and global warming.

Social anthropology, as it produces knowledge about the similarities and differences between humans and other animals, can take these pathogens crossing barriers between species as a starting point for an inquiry about transformations in our relationships with nonhuman animals. The connection between human/animal relationships and public health measures is twofold: new relations between humans and animals (such as the increase of livestock for human consumption) has produced new risks of emergence, but the techniques to mitigate these risks (such as the massive culling of poultry or the use of sentinel chickens) have also changed the way humans relate to other animals.

This book is based on an ethnographic research conducted in Hong

Kong, Taiwan, and Singapore between 2007 and 2013.[4] As these three territories have been affected by the SARS crisis in 2003, they invested in techniques to prepare for an influenza pandemic. Hong Kong was my main site of research because, being the location where the last flu pandemic had officially started in 1968, it had been equipped to detect the next pandemic virus among birds. But these three territories were concerned with an avian influenza virus coming from China, where the number of domestic poultry had dramatically increased in the last forty years. Hong Kong, Taiwan, and Singapore are three hubs for the Chinese diaspora, who could identify with the migratory birds accused of spreading influenza across the globe. One of the arguments of the book is that these three territories on the borders of China and in a distanced connection to Australia found with avian influenza a language to talk about the problems they have with mainland China, considered as an emerging power whose conditions of life and emerging threats lacked transparency. In these three settings, microbiologists have allied with veterinarians and birdwatchers to follow the mutations of flu viruses between wild birds, domestic poultry, and humans. I have increasingly spent more time with birdwatchers because I was intrigued by a question: can we see pathogens from the perspective of birds themselves? I thus came to share birdwatchers' passion for bird species and microbiologists' curiosity for viral mutations rather than becoming versed in Chinese genealogies and kinship systems, because I found in viruses a way to enter into the relations between humans and birds in the geopolitical context connecting China, Hong Kong, Taiwan, and Singapore.

In 2003, in the aftermath of the SARS crisis, three microbiologists working at Hong Kong University wrote, "The studies on the ecology of influenza led in Hong Kong in the 1970s, in which Hong Kong acted as a sentinel post for influenza, indicated that it was possible, for the first time, to do preparedness for flu on the avian level."[5] This quotation has provided the impulse for the reflection developed in this book. What does it mean to practice preparedness at the animal level? How does it differ from doing it at the human level? What does it change in the relations between humans and other animals? Is there something specifically Asian in the way preparedness has been implemented? And what can we learn about the way Asian societies have practiced preparedness "at the avian level"?[6] In short: what do "avian reservoirs" reveal for anthropologists working in Asia? Or: what do birds with flu viruses reveal about the position of Asia in the global economy?[7]

The notion of "avian reservoir" could be criticized for suggesting that

Asian populations live in too much proximity with their chickens and pigs[8]; and indeed, "avian reservoirs" sounds like a stigmatization of "Asian people" as a "reservoir for viruses" in a new version of what Claude Lévi-Strauss called "les tropiques bondés," by contrast with the "tropiques vides" of Amazonian forests.[9] But I want to take a cynegetic view of avian reservoirs to conceive them precisely as an Amazonian forest—that is, as a space where human and nonhuman animals are connected by invisible entities called "microbes" that can be captured, classified, and mapped. I will show that the notion of "avian reservoir" involves a mix of techniques I will call "pastoralist," in that they monitor birds as sheep in a flock, and techniques I will call "cynegetic," in that they follow birds as prey in the wild.

In this book, I want to reflect on the alliance between microbiologists and birdwatchers using concepts from the anthropology of hunter-gatherers. As most of my inquiry was made with ornithologists and microbiologists, I decided that the complementarity between their two perspectives on bird flu and their difference with other actors of pandemic preparedness would become my object of research. What does it change to take seriously the idea that microbiologists are "virus hunters" and "collectors of samples"? How is the imperative to be prepared for an influenza pandemic embedded in the practices of microbiologists and birdwatchers when they see relations between humans and birds through the pathogens they share in common? The anthropology of hunter-gatherer societies has shown that these groups have developed a capacity to perceive the environment through the eyes of the animals they prey on. Microbiologists and birdwatchers refuse to kill the birds they observe, or defer the moment of killing, because they need to catch something of their perspective on the environment. In contrast, public health management of the threat of avian influenza involves killing birds to protect humans without taking on their perspective; for public health officials, bird diseases are signals that something has gone wrong in the world and requires human intervention. These two different perceptions of animals' death can be called "preparedness" and "prevention." Most of the book is dedicated to clarifying this distinction.

This book thus combines a theoretical argument in social anthropology with an ethnography of human/animal relationships and public health techniques, to describe the surveillance of avian reservoirs in specific territories in Asia. It is divided into two parts: one is more theoretical and discusses the stakes of preparedness for social anthropology, while the other is more empirical and describes relations between humans and birds engaged in specific techniques of preparedness. In part I, I reflect on my position as an anthropologist trained in the French structuralist tradition, working with microbiologists in a European project and with curators in a museum. In part II, I describe my observations in Hong Kong, Singapore, and Taiwan referring to the anthropology of hunter-gatherers.

Chapter 1 discusses how anthropology has referred to animal diseases in order to think about the social. It shows that the conceptual apparatus of social anthropology, which has historically relied on the distinction between nature and society to build concepts of causality reflecting modes of intervention, has changed in parallel with transformations in the public health management of animal diseases. Claude Lévi-Strauss's diagnosis of mad cow disease is read as a more ecological approach to animal diseases, based on the techniques of anticipation of hunter-gatherers, than the views of anthropologists from Herbert Spencer to Émile Durkheim, borrowed from the observations of pastoralist societies. This chapter clarifies historically and genealogically the distinctions among prevention, precaution, and preparedness that have been used to diagnose the emergence of bird flu in Europe.

Chapter 2 looks at a recent controversy on mutant flu viruses to raise questions on the linguistic slippages of microbiologists when they are dealing with unstable entities, hypotheses, and models. Following the discussions between virologists and epidemiologists on the adequate techniques to prepare for a pandemic, it raises questions on the possibility to anticipate in the laboratory viral mutations that will emerge in nature. The notion of lure allows me to connect biosecurity concerns with techniques from hunter-gatherers.

Chapter 3 describes prevention and preparedness as different techniques to conserve the past in order to anticipate the future. It relates the emergence of virology and ornithology to the places where samples are accumulated and classified. It then traces the role of anthropology in museums where cultural artifacts are conserved to reflect on the possible interactions between microbiologists, birdwatchers, and anthropologists in the field. It also asks questions about the position of China as an empty space in the global collections of museums.

In these three theoretical chapters, preparedness is thus described as a mode of causality (justifying governmental interventions), a technique of language (connecting nature and the laboratory), and a form of visibility (producing accumulation and classification). In the next, more ethnographic chapters, I describe three techniques of preparedness as they are implemented in Hong Kong, Singapore, and Taiwan. Each of these territories provides me with the vignette that opens the chapter, which leads me to speculate how far each one of these territories preparing for disasters coming from China could be best described through each of these three terms: Hong Kong as a sentinel post, Singapore as a technological space for simulation, Taiwan as a storing repository.

In chapter 4, on sentinels, I show that relations between self and other are configured in sites where early warning signals are produced at different levels: the globe (environmental sentinel), the sovereign territory (sentinel post), the farm (sentinel chicken), and the organism (sentinel cells). In these different settings, I ask how sentinels can fail or be lured and how early warning signals are produced in situations of uncertainty, relying on ornithologists' views on sentinel behaviors. Starting from the mobilization of Hong Kong to prepare for a bird flu pandemic, I ask what it means to be a successful sentinel and what is the cost of this mode of signaling.

In chapter 5, on simulation, I analyze the performance of public health actors enacting scenarios of the coming pandemic. Asking how animals can be included in these scenarios and how these simulations become digitalized, I discuss notions of ritual, performance, play, and fiction, taking seriously the idea that virologists and bird watchers act as contemporary hunters-gatherers. The argument of this chapter is that scenarios of a bird flu pandemic allow actors to play with human/animal relationships in a reverse mode, thus anticipating their future uncertainties.

In chapter 6, on storage, I look at forms of accumulation (antivirals and vaccines) to prepare for pandemics, and I explore ethnographically their distinction with more classical forms of storage. I rely on anthropological debates about gift and exchange to cast light on the production of value in the world of microbiologists and birdwatchers. This chapter argues that the accumulation of samples and vaccines mixes preparedness and prevention, producing ambivalent debates about precaution, sovereignty, and equity.

In a style that can be called philosophical-anthropological (or fieldwork in philosophy),[10] I consider prevention and preparedness as concepts and not just as techniques—that is, I extract them from the spaces in which they can be observed to generalize them as modes of relations between humans and their environment, which I think of as cynegetic and pastoralist; but I don't want to consider these ideal types as abstract essences, and I describe how they can be mixed in actual public health practices. Similarly, I don't engage with the anthropologists of hunter-gatherers and pastoralist societies ethnographically, for this would require an attention to the diversities of forms of life that is outside the scope of this book, but

I borrow from them concepts of myth, ritual, and exchange that I use to describe contemporary techniques of preparedness. However, this doesn't mean that I refer to hunter-gatherer societies as a literary metaphor or as a romantic worldview to think about relations between humans and birds; rather, I take as seriously the ontological claims of virus hunters and birdwatchers in China as ethnographers do when they study hunter-gatherers in Siberia, Amazonia, Africa, or Melanesia.

My attraction for concepts as tools to capture relations between humans and their environment (an attraction I may share with hunter-gatherers) leads me to valorize triadic relations between concepts (a view I may share with pastoralist societies).[11] However, this doesn't mean that these concepts work in a dialectic to produce a Hegelian synthesis; nor do they correspond with each other in a systematic framework. The aim of conceptual distinctions is to do a critical work—that is, to make a difference in debates that are often confused about pandemic preparedness and thus open alternatives to securitizing views of relations between humans and their environment. Anthropology's main distinction between hunting and pastoralist societies on the threshold of domestication allows me to be critical when observing contemporary relations between humans and animals as they are engaged in pandemic preparedness. In part I, I define prevention (which can also be named "securitization") as the management and control of populations in a territory through the use of statistics, and preparedness (which can also be named "mitigating") as the imaginary enactment of disasters in a community where humans take the perspective of nonhumans.[12] I then define precaution as a mix between prevention and preparedness, since it is an injunction to protect oneself when the state doesn't control a defined territory. In part II, I show that sentinels, simulations, and storage, considered as cynegetic techniques of preparedness and described in the three ethnographic sites of Hong Kong, Singapore, and Taiwan, can also be described as pastoralist techniques of prevention when they are conceived as sacrifice, scenarios, and stockpiling. If this book may be summarized through the distinction among three P (prevention, precaution, preparedness) and three S (sentinel, simulation, stockpiling), it doesn't mean that these three terms follow each other dialectically; rather, the two P's divide each of the three S's in a diabolic mode that reflects the subversive potential of avian reservoirs.

animal diseases

PART I

CHAPTER ONE

culling, vaccinating, and monitoring contagious animals

The most visible policy in case of a zoonotic outbreak is the massive killing of potentially infected animals—technically called "culling," which involves a separation between "proper" and "improper" animals. But there are two other key techniques to control the spread of diseases among animals: the use of vaccines to produce immunity and the monitoring of mutations of pathogens through the collection of data.[1] In this chapter, I will look at how social anthropology has justified these three techniques by different views of collectives gathering humans and nonhumans. This chapter thus proposes a parallel genealogy of social anthropology and animal health management to clarify the distinction between prevention and preparedness. Four main authors in the history of social anthropology will be read in regard to four main animal diseases in the history of public health in Europe to question how the very idea of the social was challenged by animal diseases.

Culling, vaccinating, and monitoring animals assume different understandings of microbes as invisible beings emerging in relations between humans and animals and different views of the causality by which humans, animals, and microbes interact—what I will call their *forms of participation*.

One of the main questions raised under the notion of participation is how to include all actors in the management of animal diseases.[2] In the history of social anthropology, it has been used to discuss the classical notion of sacrifice. It defines a modality of causality that is not linear, as in physical causality, but holistic, in the sense that all the beings constituting a society act at the same time. This chapter will then investigate how participation as a method to include all actors in the treatment of animal diseases was also defining human collectives at the expense of animal lives. Participation, I argue, connects ideas about animal lives with assumptions about public health in an attempt to overcome tensions or contradictions in relations between humans and animals. This chapter will ask what is an illusion or a false idea in the public health management of animal diseases and how preparedness as a technique of animal health management has emerged as an attempt to overcome the contradictions met by previous techniques. If animal breeders, public health officials, and journalists are required to participate in the management of animal diseases, how can they participate following a rationality that is not sacrificial? I will argue that if prevention excludes the perspectives of animals on public health management under a sacrificial rationality underlying culling and vaccinating, preparedness includes them by extending participation through techniques of monitoring.

HERBERT SPENCER AND FOOT-AND-MOUTH DISEASE

Herbert Spencer is seldom read by social anthropologists today, but he did a lot to popularize the notion of *sociology*—a term launched by Auguste Comte in French in 1830 to define the reflexive knowledge that society takes upon itself when it encounters a public crisis.[3] The general framework of his theory can be called *evolutionist*, even if it was not influenced directly by Darwin, in that it assumed a general progress of humanity from its most primitive states to more complex forms.[4] In 1873, Spencer opened his *Study of Sociology* with the example of a breeder complaining about the public policy on animal diseases. In Spencer's account, the massive killing of sick animals is an occasion for individuals to start thinking reflexively about the society in which they live.

> Over his pipe in the village ale-house, the labourer says very positively what Parliament should do about the "foot-and-mouth disease." At the farmer's market-table, his master makes the glasses jingle as, with his fist, he em-

phasizes the assertion that he did not get half enough compensation for his slaughtered beasts during the cattle-plague. . . . Minds in which the conceptions of social actions are thus rudimentary, are also minds ready to harbour wild hopes of benefits to be achieved by administrative agencies. In each such mind, there seems to be the unexpressed postulate that every evil in a society admits of cure; and that the cure lies within the reach of law. . . . Minds left ignorant of physical causation, are unlikely to appreciate clearly, if at all, that causation so much more subtle and complex, which runs through the actions of incorporated men.[5]

Here Spencer seems to conflate, in the words he attributes to the laborer, two very different diseases: foot-and-mouth disease and cattle plague. Cattle plague (also known under its German name *rinderpest*) was the most severe animal disease in Europe in the eighteenth century. Originating in Asia, it emerged in Italy in 1711, England in 1714, and then spread, by contact and by air, throughout the European continent. It was estimated to have killed 200 million cattle in England between 1740 and 1760, after high fever and loss of appetite, triggering the formation of veterinary science at the end of the eighteenth century.[6] The disease then moved to Africa at the beginning of the twentieth century and dramatically affected the cattle in Sudan in the 1930s, when the British anthropologist Edward Evans-Pritchard was studying among the Nuer and Dinka pastoral societies.[7] The World Organization of Animal Health declared cattle plague eradicated in 2011 after the success of a global vaccination campaign.

By contrast, foot-and-mouth disease, first described in Italy in the seventeenth century, became a major concern in England in the 1870s,[8] which explains why Spencer takes it as a priority case to start his *Study of Sociology*. Foot-and-mouth disease does not kill animals, but provokes symptoms (such as fever and blisters) that make it impossible for animals to circulate and be commercialized. It became a crucial issue for Great Britain in the interwar period, when it affected the cattle imported from Argentina, and in the 2000s, when it periodically blocked the economy of the country. While cattle plague is a disease of pastoralist societies, enhanced by the agricultural revolution of the eighteenth century, foot-and-mouth disease has been called by historian Abigail Woods a "manufactured plague," as it reveals the vulnerabilities of a global interconnected economy.[9] It therefore linked veterinary science to emerging knowledge such as laboratory clinic and epidemiology.[10]

Spencer's narrative makes sense in this new situation. The laborer complains that the Parliament has slaughtered his cattle without giving him enough compensation. Foot-and-mouth disease is indeed associated with a massive "stamping out" policy. While cattle plague was soon treated with inoculation, the virus causing foot-and-mouth disease, identified by Friedrich Loeffler in 1897, is very unstable, which makes vaccination more difficult. When it appears in a flock, agricultural authorities may decide to kill the whole flock to prevent the spread, even if no animal has died from the disease. It is, therefore, difficult to explain the rationality of culling, and while compensation may soothe breeders, it never pays for the loss—psychological, financial, and genetic—of the flock.

The role of sociology, for Spencer, is therefore to analyze the "mind" of these breeders who do not understand the rationality of the society they live in, such as the laborer and his master in the scene he depicts. From Spencer's perspective, cattle breeders—portrayed as heavy drinkers—are not far from primitive tribes under the hallucination of their fetishes. Spencer discards the knowledge they may have about animal diseases as primitive superstition, following the prejudices of British elites at the time. Historian Keith Thomas recalls that the new science of observation and classification developed by eighteenth-century British urban elites made long-held popular attitudes about nature obsolete. It was then said that "when cattle had the murrain, the husbandman cut a hole in the beast's ear and inserted a root of bearfoot,"[11] assuming an analogy between the name of the plant (bearfoot) and the disease to cure (*murrain* is an old term encompassing several cattle diseases such as cattle plague and foot-and-mouth disease).

Spencer's argument about social causation is linked to his ideas about proper government. For him, cattle breeders cannot understand social causation, which is the same as physical causality but made more complex by the interactions between humans, as two modalities in the association of ideas. They expect the state to cure their diseases just as "primitives" bow to their fetish to ask for rain, says Spencer. Following a liberal tradition, Spencer argues that social causation should be modeled on physical causation, letting the complex chains of modern economy develop with restricted interventions from the state.[12] For him, social actions and mental ideas follow the same rules as natural flows and diverge only by what he describes as an excess of nervous irritation. The role of sociology is, therefore, to bring angry cattle breeders and reluctant primitive tribes to quietly accept the authoritarian interventions of the state. While ordinary individuals anticipate the future based on inferences from past observations, Spencer argues, only the state and the market can organize prevention based on statistics.

Spencer criticizes the role of what we would now call *the media* in the contagion of ideas. For him, this slight divergence between physical causation and social causation, which is produced by excessive irritation and alcohol consumption, is enhanced by the spread of ideas within society. In an empiricist tradition, Spencer examines the subjective values of testimonies on the severity of diseases. Natural biases, he says, are reproduced by journalists, who launch "false ideas" in "the public mind" that can then never be corrected or falsified. The role of sociology is, therefore, to replace the illusions of inference—which Spencer compares to the refractions of the moon on a lake—by the accumulation of data through observation, which would then be, in Spencer's view, as solid as the observations of astrologists.[13] What is more, sociology should propose a theory about the causality that produces false ideas. The proposal of such theories makes sociology, as the only "reflexive" science, superior to astronomy and physics. Since natural and social causalities are the same, with only different degrees of complexity, a sociology of the contagion of ideas must accompany a medicine of the contagion of bodies to produce a modern public health policy.[14]

This conception of the social is widely shared in the contemporary management of animal diseases. In liberal societies, culling policies appear as necessary interventions of the state to regulate the free circulation of animals for human consumption. The knowledge breeders have about their animals is considered too emotional or superstitious to be taken into account by public health authorities. Experts are supposed to have a clearer view of the order of phenomena that must be regulated through statistic reasoning and state interventions. This definition of prevention as an expert view of the future is contested today by participatory veterinary medicine, which shares interesting parallels of ecological holism with how Robertson Smith understood invisible but powerful entities in his reconstruction of ancient Semitic religions.

WILLIAM ROBERTSON SMITH AND BOVINE TUBERCULOSIS

As a professor of Semitic languages at Cambridge University, Robertson Smith was particularly influential on one of the founders of British anthropology, James Frazer, and helped him develop a theory of the sacred that departed from the evolutionist assumptions of Spencer. When Robertson Smith was writing, the most discussed animal disease was tuberculosis, endemic in cows and humans and causing massive killings of cows and less massive death among humans. Smith died of this lung disease at the age of forty-eight, and three of his brothers and sisters died of it at an early age.[15] It may be argued that this exposure to tuberculosis made Robertson Smith more sensitive than Spencer to the fate of culled animals and to think differently about infection and contagion.[16] Instead of calculating fair compensations for culling, Robertson Smith used the theological notion of sacrifice to describe how a new collective rationality emerges by investing animals with shared affects under the sacred. Sacrifice became a way to ritually form and reinforce subjective feelings about the proper social roles of animals. It may be said that Spencer required breeders to sacrifice their interests to that of society and that Robertson Smith justified this sacrifice as a spontaneous reasoning of breeders themselves. In other words, Robertson Smith exposed the sacrificial rationality that remained hidden in Spencer's liberal reasoning to justify the massive killing of sick animals.

At the end of the eighteenth century, British and German physicians were enmeshed in a controversy about the transmission of tuberculosis from cows to humans. Robert Koch, who discovered the bacterium (or bacillus) causing tuberculosis in 1882 and gave it his name, denied that it was transmitted from cows to humans. Though the symptoms were the same in both species, the shape of the bacteria appeared different under the microscope.[17] The philosopher Gottlob Frege referred to this controversy when he wrote, "Investigators who had discussed among themselves whether bovine tuberculosis is communicable to man, and had agreed finally that such communicability did not exist, would be in the same position as people who had used in the conversation the expression 'this rainbow,' and now came to see that they had not been designating anything by this word, since what each of them had had was a phenomenon of which he himself was the owner."[18]

For Frege as for Koch, the subjective fear of a transmission, which produced multiple views of the disease, should be replaced by the objective representation under the microscope.[19] The notion of "bovine tuberculosis," says Frege, appeared as contradictory and illusory as that of a rainbow. It was only in the 1920s that British physicians, who were concerned with the safety of cow's milk and meat, could prove that 20 percent of human tuberculosis came from cows. They had to show, against Koch, that the same pathogen can take different forms when it passes from animals to humans. This required a theory of microbial mutations that did not exist at that time. To prove objectively the existence of pathogenic mutations, they had to use statistics rather than microscopes. What was needed to understand tuberculosis was a science of the social and ecological milieu in which the pathogen takes new forms.

Robertson Smith made a similar demonstration in the anthropology of religion. After traveling to the Middle East to find the sources of the Bible, Robertson Smith defined sacrifice as a communion, in the etymological sense of a meal shared with the divinities.[20] While Edward Tylor, the founder of British anthropology, had defined sacrifice as a gift to the gods, Robertson Smith noticed that this definition assumed that "primitive societies" had a concept of property. He remarked that sacrifice makes distinctions between entities considered "kin" not through the categories of social property but through the supernatural powers attributed to invisible beings. We can reframe this idea stating that Smith described how the social emerges out of interactions between humans, animals, and microbes rather than presupposing the social as a framework for those interactions.

The role of sacrifice, in Smith's view, was to fix what he called "supernatural beings," which we could call microbes, since they act despite being invisible, into a holy place, which looks like the place of collective intervention by the state. Smith uses the Polynesian term *taboo* to describe precautionary measures applied to forces he qualifies as "infectious."[21] All the members of a group—or kin—participate in sacrifice because all are exposed to the supernatural forces constituting a milieu:

> In the beginning, the beasts and birds of the sanctuary, as well as its vegetation, were conceived as holy because they partook of the pervasive divine life. We may conceive the oldest sanctuaries as charged in all their parts and pertinent with a certain supernatural energy. This is the usual savage idea about things that are taboo, and even in the higher religions the process of subsuming all taboos under the conception of the holiness of the personal god is always slow and often imperfectly carried out. . . . Holiness, like taboo, is conceived as infectious, propagating itself by physical contact.[22]

By contrast with Spencer, Smith makes a distinction between physical causation and social causation. While physical causation acts through the contact of one thing with another, Smith argues, social causation acts through the integration of a thing into an order that supersedes it and constitutes its meaning. This may be related to the fact that, unlike Spencer, Smith was concerned with the infectious aspects of disease rather than with its contagious effects in society, or with the emergence of the social rather than its effect on the mind. Historian François Delaporte has clearly opposed these two medical schools of the nineteenth century: while contagionists viewed diseases as spread by contact and recommended creating boundaries, infectionists returned to the origin point of an epidemic and aimed to restore the circulation of life around that place by cleaning habitats or killing animals.[23] Smith suggests a return to the site of infection in order to describe how society emerges in this site through the sharing of affects. In this site, bodies participate in an invisible flux of forces that orients them toward a collective future. While disease is abnormal for Spencer as a deviation from the physical laws of contact, it is normal for Smith as constitutive of the social matrix of life. Indeed, tuberculosis was rampant among human and nonhuman animals at the end of the nineteenth century and revealed the unequal ways in which they occupied social habitats.[24] For Smith, infection is the site where the social emerges as different modes of exposure to disease, and sacrifice is the action through which these differences become compatible.

Robertson Smith thus appears as more liberal than Spencer when he rehabilitates the ecological knowledges and attachments that are often disregarded in authoritarian justifications of state intervention—a view that was influential on Frazer's *Golden Bough*, which collects narratives about animals and plants. For Smith, those who lived close to the holy place had an intimate knowledge of the measures of precaution that were adapted, while the economy of sacrifice and the emergence of property produced categories that could distance humans from this intimate knowledge. This view may have influenced the Chinese, Cambridge-trained physician Wu Lian-Teh, who developed an anthropological theory of the knowledge of marmot hunters during the epidemic of pneumonic plague in Manchuria in 1910–11, to explain why plague affected coolies and not hunters. "The natives have known plague for generations," he wrote. "They possess a working knowledge of the disease and adopt elaborate precautionary measures for fighting it."[25]

But as Christos Lynteris has shown,[26] theories of local knowledge of the animal origins of plague multiplied in the years before 1910 and produced contradictory visions of the transmission of plague between marmots and humans, for which Wu Lien-Teh proposed an unstable compromise. Indeed, the contradiction in the notion of zoonosis noted by Frege cannot be solved through the concept of local knowledge: how can the same pathogen transmit between different animal species? How can individuals participate in social life if it involves such contradictory beings as zoonotic pathogens? Smith aimed to build an objective science of the subjective exposure to disease through the concept of sacrifice, but he failed to describe the differences of exposure because he lacked knowledge about the mutations of pathogens. Such knowledge came with vaccination, which opened new ways to conceive the social.

EMILE DURKHEIM AND SMALLPOX

Émile Durkheim, who founded sociology as an empirical science after Comte had assigned its place within other sciences, famously read Robertson Smith's *Lectures on the Religions of the Semites* as a "revelation."[27] This reading is often attributed to Durkheim's turn from a sociology of economy to a sociology of religion. Yet Durkheim criticized Smith through an argument he borrowed from the sociology of law, which commands his whole theoretical enterprise. Law qualifies both the probabilistic reasoning defended by Spencer and the local knowledge valorized by Smith as ways to mitigate the intervention of the state. Following Durkheim, who took vaccination as a model to conceive the social, precaution is not only a local knowledge but a technique of government, because the state must be precautious when using the vaccine as a technique to intervene in social life.

While Smith drew an evolutionary genesis of property through sacrifice, Durkheim refused the idea that primitive societies ignored property and considered sacrifice to be the first instantiation of property. Smith makes a distinction between sacrifice as communion in the holy flesh of the animal, which historically came first in his view, and sacrifice as expiation of sins toward the sacred body of the god, which came second. For Durkheim, by contrast, the idea of a sin against property is the first moral reasoning of human societies, and all other notions of social life are engendered by the delimitations it creates between what is proper and what is improper. Durkheim thus starts where Robertson Smith ends, with the notion of the sacred as that which attracts and repels at the same time, because he considers the intervention of the state as the origin of social life, but also because he has a different reading of the totemic sacrifice. For Smith, who relies on McLennan's systematization of the Polynesian notion of taboo, sacrifice starts with precautions when handling the sacred. For Durkheim, who borrowed the description of Australian rituals from Spencer and Gillen, sacrifice begins with a delimitation of the social space where natural beings become emblems or symbols of the group. While for Smith, the holy place—or sanctuary—is open to fluxes of life, for Durkheim the sacred space is defined and bounded; it is a church or a court that categorizes social life according to its degrees of purity.

Durkheim thus proposes to explain what Robertson Smith has only described: the ambivalence of the sacred. How can the sacred be at the same time attractive and dangerous, pure and impure? For Durkheim, this apparent contradiction in social entities is due to the fact that collective life creates a new form of causality, which is the mental capacity to make distinctions between things. Durkheim uses two metaphors to describe the emergence and action of collective consciousness: chemical (effervescence) and biological (contagion). As in a chemical reaction, the interaction between living beings creates an entity that is more than the sum of its parts, so that Durkheim characterizes the emotional state of individuals during sacrifice as a "collective effervescence." But he also notes that these new forces appear to individuals as a contagious flow of invisible beings. Individuals approaching the space of collective consciousness are touched by a "virulent force" that covers them with impurity and are then redeemed by a sanctifying power through ritual.[28]

Durkheim borrows his metaphors from the new medical rationality instituted by Louis Pasteur, with whom he shares the will to found science on a new concept of causation that could cure diseases.[29] First working as a chemist and then as a biologist, Pasteur discovered the mechanisms of attenuation of the virulence of microbes while experimenting on fermentation and then on vaccination. He showed that a shift in the structure of biological molecules could change their properties.[30] In particular, he explained how the transmission of a microbe from one species to another, such as the inoculation of cowpox in humans, could modify its virulence and trigger an adapted immune response. While the principle of vaccination—the use of a pox virus from cows to produce antibodies for smallpox—had been empirically discovered by Jenner, it is thus rationally explained by Pasteur as an attenuation of microbes' virulence. In the same way, Robertson Smith discovered empirically that the sacred either destroys or sanctifies those who approach it, but Durkheim claims to explain it rationally by the spatial configurations of social life.

Durkheim mentions Pasteur's experimentations on smallpox in a passage of his *Rules of Sociological Method*, where he discusses the distinction between the normal and the pathological. Durkheim famously argues that for a sociologist, there is no intrinsically pathological phenomenon: what appears as pathological from one perspective, such as a criminal act or a religious ceremony, can be seen as normal from another perspective. Durkheim goes as far as to say that a certain dose of disease is necessary for the dynamics of social life, because it shows that social life does not follow the mechanical laws of physical causality. This is where he takes the example of smallpox:

> Sickness does not always leave us at a loss, not knowing what to do, in an irremediable state of inadaptability; it merely obliges us to adapt ourselves differently from most of our fellows. Who is there to say that some sicknesses even exist which in the end are not useful to us? Smallpox, a vaccine of which we use to inoculate ourselves, is a true disease that we give ourselves voluntarily, yet it increases our chance of survival. There may be other cases where the damage caused by the sickness is insignificant compared with the immunities that it confers upon us.[31]

If Pasteur's discovery is that microbes can artificially be attenuated to produce a memory of the body, Durkheim also conceives the social as a form of disease that instantiates a collective memory. For both of them, interactions between cells or individuals produce a collective rationality that is memorized by all of them to protect themselves against future encounters. Pasteur and Durkheim share a republican faith in the capacity of the state to intervene in the collective body for its sake, because these interventions, even if they cause mild diseases, protect against major diseases by creating a specific memory of the organism.

This view contrasts with the free-trade approach shared by Spencer and Smith, because it leaves space for a differentiated state intervention. In the light of the debate on animal diseases, we see that the disagreements between Durkheim and Smith are both political and ontological. If the sacred is infectious, it means that its cause cannot be known in itself, but only in the multiple life-forms that express it. If it is contagious, by contrast, it means that its cause is clearly known as a first-order causality acting on the lives of individual bodies. Durkheim goes further than Spencer in that he conceives the intervention of the state not through the form of law that obliges apparently irrational breeders to kill all their cattle (as in Spencer's view), but through norms of immunity that adapt vaccination to the specificities of animals. In that sense, Durkheim gave Pasteur the sociological tools to pasteurize France and the rest of the world by describing the social as various forms of life produced by shared norms.[32]

In the year 1880, when Robertson Smith was conceiving his theory of religion, outbreaks of tuberculosis in cattle forced the British state to kill hundreds of cows. A group of experts was consequently appointed to classify meat by its exposure to infectious diseases and not only by its risks of alteration.[33] In 1912, when Durkheim published his *Elementary Forms of Religious Life*, the physician Albert Calmette and the veterinarian Camille Guérin successfully attenuated the Koch bacillus and started a massive vaccination of cows with the BCG (Bacille Calmette-Guérin). This would change the relations between the French state and cattle breeders in the interwar years, since the "social problem" of tuberculosis seemed to be solved by a massive intervention of the state through vaccination.[34] As a precautionary measure, vaccination is more rational than culling, because it is adapted to the population. And even if vaccination has side effects, those who die of vaccination are supposed to sacrifice for the protection of the whole population against a disease that potentially affects all individuals.

Yet the intellectual debate on Durkheimian sociology in the interwar period can be paralleled to the failures of vaccination. In the interwar period, massive vaccination of people living in France's countryside and colonies, who were often used as experimental subjects to test the effects of vaccines, led to a series of casualties.[35] Lucien Lévy-Bruhl, a distant member of the Durkheimian school, mentions these accidents at the end of his popular book *Primitive Mentality* in 1922. While colonial doctors were surprised when the "natives" asked for money after they were vaccinated, Lévy-Bruhl explained that for them, every medical intervention is an accident that must be compensated at the level of supernatural beings, where it finds a moral formulation in terms of collective obligations.[36] Lévy-Bruhl thus uses Robertson Smith's vocabulary of participation to rehabilitate the local perception of events against the state's representation of the laws of nature. Accidents, he argues, are never perceived as natural causes but as opportunities for supernatural beings to appear; hence the role of shamans or medicine men, who train individuals how to encounter invisible beings. In Lévy-Bruhl's description, natives are constantly prepared for the emergence of supernatural beings and don't rely on the state's boundaries to protect them from impurity. While smallpox was used as a technique to extend the protection of European states to the whole globe as a definition of modernity, Lévy-Bruhl's view resonates with the concerns for emerging infectious diseases after the success of the eradication of smallpox in the 1970s. In a world where pathogens emerge from changing relations between humans and animals, public health officials may learn from "primitive" techniques to think about the "supernatural" beings emerging from these relations. Every new campaign of vaccination becomes a test of the capacity to react to a new pathogen, without the possibility of relying on a collective body such as society to mitigate this uncertainty. All individuals become guinea pigs for new campaigns of vaccination as the natives of the colonies at the beginning of massive vaccination.

Henri Bergson, discussing Lévy-Bruhl in *The Two Sources of Morality and Religion*, came to this conclusion in 1932. What does it mean, asks Bergson, to say that natives ignore hazard? We think we know what hazard means because we assume there are large series of numbers built by statisticians who calculate the probabilities of accidents. But in fact, when accidents occur, ordinary people behave like natives: they attribute the event to an invisible entity and explain its human significance by projecting an intention onto natural beings.[37] Bergson takes the example of the hunter, who knows the mechanical laws that will lead his arrow into the prey, but still addresses the spirit of the prey through myth and ritual. The hunter, says Bergson, thus reduces the margin of uncertainty around the trajectory of the arrow in the same way modern individuals think when they perceive a disaster—for instance, an earthquake or a declaration of war—as if it had always been there. For Bergson, imagination creates virtual entities within the perception of the world to compensate for the discouragement of intelligence when it calculates the chances of action.[38] Bergson calls it an intentional causality because it attributes intentions to things so that it is possible to act with them as if they were persons—what could be called a "virtual space of simulation." It thus becomes possible to use techniques of prevention based on risk calculation such as culling and vaccinating while using techniques of preparedness to simulate disasters.

From Spencer to Durkheim to Bergson, we have moved from prevention to precaution to preparedness in the management of animal diseases. These involve different conceptions of social causality. While Spencer conceives social causality on the model of physical causality, which makes it possible to prevent animal diseases based on probabilistic knowledge of nature, Durkheim argues with Robertson Smith that social causality is entirely different from physical causality, as it is based on the action of the state protecting the individual through collective reflection or symbols. Lévy-Bruhl and Bergson, returning to Robertson Smith's description of the affective aspects of participation, described the virtual spaces in which individuals prepare themselves for future catastrophes through the imagination of intentions within natural phenomena. Yet they missed the concept of *sign* to describe how the future is already inscribed in the perception of present events. In French anthropology, this concept was introduced by Claude Lévi-Strauss, who redefined social causality based on structural linguistics.

CLAUDE LÉVI-STRAUSS AND MAD COW DISEASE

Claude Lévi-Strauss has praised Bergson's philosophy for describing totemism "from inside" after Durkheim had described it "from outside."[39] What may appear as an ironic homage is in fact a strong endorsement of Bergson's conception of language. Bergson, says Lévi-Strauss, anticipated the structuralist view of totemism as a use of zoological or botanical classifications to inscribe future events in already existing systems of signs. At the end of the 1990s, Lévi-Strauss applied this conception to the social and anthropological aspects of mad cow disease. Rather than arguing for prevention or precaution to justify risk management, Lévi-Strauss showed that "mad cows" signaled a catastrophe for which humans have to prepare: the end of the industrial system of meat production and the return to hunters' relations between human and nonhuman animals.

On November 24, 1996, Lévi-Strauss published a short article in the Italian journal *La Repubblica* entitled "La mucca è pazza e un po' cannibal" (The cow is mad and a little cannibal). He commented upon the recent declaration by the British government that a series of cases of Creutzfeldt-Jakob disease, particularly among young people, was linked to the consumption of meat coming from cows infected with spongiform encephalopathy. This disease, known by veterinarians since the eighteenth century to affect sheep under the name *scrapie*, had spread within cattle over the previous twenty years due to the change in the cooking temperature of the meat and bone meals given to cows. While the consumption of beef fell in Great Britain, and while European countries imposed embargos on British beef, journalists started to talk about "mad cows" to describe the symptoms caused by nervous degeneracy and about "cannibal cows" to explain that the disease was caused by the industrial mode of production forcing them to consume beef as feed.[40]

Lévi-Strauss had known about this disease since the 1960s, when he had followed the debate about *kuru* in Papua New Guinea. At the beginning of the 1950s, Australian colonial officers in Papua New Guinea reported about a mysterious nervous disease that was transmitted among women in the Foré tribe of Papua New Guinea. Ronald and Catherine Berndt, two anthropologists working among the Foré, related the psychosomatic symptoms of kuru (delirium, trance, choking) to witchcraft practices. Carleton Gajdusek and Vincent Zigas, a microbiologist and a physician who reported about the disease in medical journals in 1957, argued that kuru was a neurodegenerative disease that could be genetically transmitted. Robert Glasse and Shirley Lindenbaum, two anthropologists, showed in 1961 that the mode of transmission of kuru victims was not genetic but social, as it involved funerary practices of women and children eating the brains of deceased persons. If kuru was an emerging infectious disease, it was then possible to find its causative agent and understand how it thrived in a new ecology. Gajdusek successfully collected brain samples from victims and transmitted them to monkeys, for which he earned the Nobel Prize in 1976. However, he failed to find what he called the "slow virus" that would explain the nervous degeneration, leaving open a path that Stanley Prusiner was to complete by showing that the disease was caused by a shifting protein he called a prion.[41]

In the *Courrier de l'UNESCO* in 1961, Lévi-Strauss wrote that kuru was "among these imported diseases against which natives have no immunity" and was caused by "the mysterious remains of a civilization that triggered it without introducing it."[42] In 1968, he published an article by Robert Glasse in *L'Homme* that he presented as "a symbolic interpretation of cannibalism."[43] The problem for Lévi-Strauss was to understand how the expansion and reduction of cannibalistic practices among the Foré under the pressure of encounter with the whites had provided the conditions for the development of kuru. A symbolic interpretation would insert cannibalism into a set of rules regulating proximity and distance in a society. Lévi-Strauss's theory of cannibalism can be compared to his theory of incest, since these are two prohibitions, one in the domain of food and the other in the domain of sexuality, that condition in his views all the rules of gastronomy and kinship. Following Durkheimian views on vaccination, Lévi-Strauss saw kuru as revealing chains of signaling and social relations through which humans can anticipate the future.

After reflecting on the exocannibalism of the Tupinamba in *Tristes tropiques*, Lévi-Strauss dedicated his course at the Collège de France in 1974 to relations between "cannibalism and ritual cross-dressing," looking at cases of endocannibalism from Papua New Guinea. In these lessons, he rejected the view that cannibalism is motivated by an aggressive drive and asserted instead that it is a regulated practice through which humans communicate with each other. He compared this regulated human communication to the communication between cells. Cannibalism, according to Lévi-Strauss, is only the limit of a gradient of identifications of the self to the other, from communication to sociability, predation, and incorporation. Rather than restricting cannibalism to a few cases whose authenticity is always discussed, Lévi-Strauss proposes a large view of cannibalism as the "spark" or the paradoxical situation that gives rise to a range of relations between self and other. Hence the title of his 1993 article in *La Repubblica*, "Siamo tutti cannibali" ("We Are All Cannibals").[44] In the same way that Lévi-Strauss showed in *Elementary Structures of Kinship* that every society is somewhat incestuous in that it does not completely follow the rules of generalized exchange, but always restricts alliance to a limited set of partners, he shows in his Mythologiques that cooking is never entirely cultural but assumes a natural dimension through the consumption of meat, which reminds humans of the origin of cooking in the killing of animals.

Lévi-Strauss thus produces a complete reversal of the mediatic expression "cannibal cows." It does not mean that cows have been denatured by the food industry, but on the contrary that they have become humanized, integrated into a species that defines itself by its capacity to absorb the other.[45] Following a utopian proposal from Auguste Comte, Lévi-Strauss hypothesizes that mad cow disease will lead to a distinction between two kinds of cows. Some, considered to be meat-producing machines, would return to a vegetarian diet and to a state of wilderness where they would be hunted by humans. Humans would eat them with the same respect, Lévi-Strauss suggested, that cannibals express when they ingest their ancestors or their enemies. Others, still being fed with animal proteins, would be assigned "watching over energy sources, and machines could be entrusted to them."[46] They would work as sentinels for the diseases circulating in the cows returned to wildness. In this visionary text, Lévi-Strauss anticipated the legislation requiring that in European slaughterhouses, the brains of cows should be tested for the presence of prions, thus forcing workers to bow upon the skulls of cows as a kind of funerary homage similar to those of the Foré.

What does it mean to consider cannibal cows as "servants of humanity" tasked with monitoring cattle returned to the wild? How does it offer a new perspective on the relations between humans and animals at a time of concern for the spread of zoonoses in the European food industry? Lévi-Strauss used an apocalyptic tone to state that mad cow disease and other zoonoses signaled the dawn of an age when the consumption of meat would decrease while the number of humans on the planet keeps increasing. His argument was that humanity should prepare itself for this coming catastrophe by communicating with animals through warning signals. Lévi-Strauss thus looks at mad cow disease from the "cannibal" perspective of the Foré people, which he related to what he had observed in Amazonia in the 1930s. It would be misleading to think that Lévi-Strauss considered animals as "subjects" because he showed empathy for the suffering of animals killed for meat, as this would endorse a Western separation between nature and culture. From an Amazonian or Melanesian perspective, translated in structuralist or poststructuralist method, microbes common to humans and animals are the real entities, and the separations made by humans to mitigate their threats are social constructions.[47]

Philippe Descola has showed that the structuralist method develops a new conception of social causality as the emergence of collective properties following an event whose meaning varies depending on the ecological milieu.[48] Rather than starting from the boundaries of a territory transgressed by a pathogen, he suggests starting from the universal fear of eating poisoned meat and seeing how it distributes intentionalities and physicalities, or invisible and visible entities, in different societies.[49] In a similar fashion, I have proposed to distinguish modes of anticipation of the future that are often conflated, prevention, precaution, and preparedness, by linking them to techniques of government by diseases and modes of relations between humans and animals. In Spencer's view, which is commonly shared as a liberal version of public health, the state lets natural causality regulate animal livestock and only acts through stamping out, which doesn't assume a separation between natural and social causality. It can therefore be described as "analogistic" in the sense given by Descola—that is, as an attempt to regulate the proliferation of beings through a sovereign gesture of sacrifice. In Durkheimian sociology, the intervention of the state adds a level of causality that is properly social by distinction with natural contagion, because it relies on the memory of previous contagions by vaccination. It mobilizes precaution to define a space of risk assessment for the state and its experts in a way that Descola would characterize as naturalistic (although it often conceives of itself as totemic). In Lévi-Strauss's anthropology, the social is conceived as a series of signs produced in a situation of communication between hunter and prey. The surveillance of sentinel animals follows the paths of contagion across borders between species. Descola calls this view *animistic*, a term that has taken in the anthropological debate a meaning that is different from the one it had at the time of Spencer and Tylor because it relies on the perception of signs in daily interactions rather than on the imagination of spirits in extraordinary situations.

This genealogy of social sciences has allowed me to distinguish different views of the relations between humans, animals, and microbes among contemporary experts who have to manage animal diseases. In the follow-

TABLE 1.1

ANTHROPOLOGISTS	**Spencer**	**Robertson Smith Durkheim**	**Lévi-Strauss**
ANIMAL DISEASES	Foot-and-mouth disease Cattle plague	Tuberculosis Smallpox	Mad cow disease Bird flu
RATIONALITIES OF RISK	Prevention	Precaution	Preparedness
MODES OF INTERVENTION	Culling	Vaccination	Monitoring
CAUSALITY	Natural	Social	Structural
MENTALITY	Ideas/reflection	Emotions/ participation	Signs/imagination
RELATIONS BETWEEN HUMANS AND ANIMALS	Sacrifice	Sacred/ supernatural	Sentinels

ing chapter, I will use these distinctions to clarify a controversy about risk management for avian influenza. I will show how microbiologists can be conceived as following an "animistic ontology" or "cynegetic techniques" when they collect samples to monitor pathogens to anticipate a pandemic, and as following an "analogistic ontology" or "pastoralist techniques" when they produce statistics on the paths of contagion to justify the intervention of the state. The conflation between these ontologies produces a debate about the need for precautionary measures to mitigate the risks of biological research for society at large, using naturalistic oppositions between nature and society. The tension between these two modes of government comes from the instability of the entities they try to stabilize: pathogens crossing borders between species. In this chapter, I have argued historically that prevention and preparedness as techniques of government are two consistent modes of stabilization of these entities, which enter in tension in the discussions on precaution. The next chapters will display the consistency of these techniques when they are faced with contemporary animal diseases.

CHAPTER TWO

biosecurity concerns and the surveillance of zoonoses

"RED HERRING"

In November 2013, I was taking part in a meeting of the consortium Antigone (Anticipating the Global Onset of Novel Epidemics) at the Pasteur Institute in Paris. I was the only anthropologist in a network of microbiologists, funded by the European Commission, that was seeking to unfold the drivers by which pathogens (viruses and bacteria) cross barriers between species.[1] My role was to produce statistical data, through surveys and questionnaires, on the social and cultural factors that favored the transmission of microbes from animals to humans, in addition to the natural mechanisms of mutation and selection.[2] But taking part in these meetings in various European cities allowed me to reflect on the way this world of Big Science and Big Data produces knowledge about small beings.[3] Excitingly, I found, observing this knowledge in the making blurred the very distinctions between nature and society that presided over the assumed division of labor between biologists and anthropologists.

I was struck by the presentation of Ron Fouchier, a researcher in virology at the Erasmus Medical Centre in Rotterdam. This research was controversial, and one of the goals of the presentation was to know if the Antigone consortium would share the credit and responsibilities attached to its investigator's name.[4] Fouchier was famous in the world of virologists for modifying the avian influenza virus in such a way that it became transmissible through air between mammals. Avian influenza had been one of the models for studying the spread of infectious diseases from animals to humans since the emergence of the H5N1 virus, transmitted from birds to humans in Hong Kong in 1997.[5] A global alert had been raised in the years 2003–5 about the pandemic threat of this new virus as it moved from Asia to Europe and Africa. This mobilization found its climax in the 2009 global fight against H1N1, a virus that was transmitted from pigs to humans in Mexico and spread rapidly to the rest of the world. In March 2013, a few months before Fouchier's presentation in Paris, a new H7N9 avian influenza virus had emerged in Shanghai, killing around one out of three humans reported as infected. By contrast, H5N1 killed two out of three humans it infected, while the 2009 H1N1, so highly contagious that it became impossible to count the number of persons infected, was less lethal than the seasonal flu virus.[6]

In this narrative of the emergence of new influenza viruses, "H" and "N" refer to the proteins commanding the entrance and the release of the virus in the cell of the host-organism, and the number indicates the chronological order of emergence. Hence H1N1, emerging in 1918 among humans and killing probably fifty million of them, is considered the first pandemic flu virus of the twentieth century, because it was both lethal and contagious. Fouchier's research aimed at answering a set of fundamental questions, with huge implications for public health: how to explain that flu viruses, which trigger ordinary reactions when they pass between humans, have devastating effects when they are transmitted from animals to humans? And could the H5N1 or the H7N9 viruses spread between humans as fast as the pandemic H1N1 virus while maintaining their high rates of lethality? The answer to that question seemed to lie in the molecular analysis of these viruses; but it was also necessary to see how these molecular mechanisms reacted to given environments—particularly inside and outside the lab.

In a few slides, Fouchier explained that five nucleotides separated the "wild type" of H5N1, found in Chinese farms and markets, from the "mutant" H5N1 he had designed in his lab in Rotterdam. He concluded that these five nucleotides brought a pandemic potential to a virus that was already known for its high lethality. When the representative of the Ethics Board of the Antigone consortium asked him, "What are the implications of your research for surveillance programs?," Fouchier answered, "If the Chinese find a flu virus that has three or four of these mutations, they can sound alarm. It is a red herring." All the biologists in the room laughed; I was the only one to miss the joke.

A red herring is defined by the *Oxford English Dictionary* as a logical fallacy that misleads or distracts from important issues. Its primitive meaning may have referred to the use of a smoked fish to divert hounds from the correct route when hunting. What Fouchier meant to say, I was told by biologists during the break, was that the mutant H5N1 would be a "red alarm" that would provide early warning signals of an emerging flu pandemic. That term shares the name of a famous video game that simulates shooting a virtual target. Fouchier had, wittingly or not, substituted the term *red herring*, which made it seem like he was criticizing his own research, for "red alarm," which was an acceptable justification of his work. I later learned that Jeremy Farrar, director of the Oxford University Clinical Research Unit in Ho Chi Minh City and member of the Scientific Advisory Board of the Antigone consortium, had publicly criticized Fouchier's research as a "red herring."[7] Jeremy Farrar, who later headed the Wellcome Trust, had a personal reflection on how to communicate about the imaginary potentials of pandemic preparedness. He received Lêna Bùi as an artist in residence at his clinic and during his epidemiological investigations in rural Vietnam in 2012. He then displayed at the Wellcome Collection the film she made on the intimacies of poultry working and the material presence of poultry feathers in the atmosphere of villages.[8]

Fouchier seemed to be familiar with this kind of ambiguous statement. The first time he exposed his research on the transmissibility of H5N1 between mammals at a flu conference in Malta in September 2011, he said, "I have done something very stupid." This sentence was often quoted in the ensuing debate over whether Fouchier's research followed the rules of biosecurity for manipulating dangerous pathogens. But he later justified himself by saying that the Dutch word for "stupid" can also mean "simple." Just as flu viruses mutate when they cross species barriers, it seemed that words slipped from one meaning to another when they shifted between languages and venues.

Indeed, Fouchier's technique was very rudimentary. He had inoculated a ferret with the H5N1, passed its nasal swab to another ferret, and repeated this procedure ten times. He then built a cage in which an infected ferret was isolated from a healthy ferret in the adjacent cage in such a way that there could only be air contact between them. He showed that the infected ferret's sneeze had transmitted the H5N1 virus to the other ferret, and then he sequenced this "mutant" flu virus.

His results converged with researches led by Yoshi Kawaoka at the University of Wisconsin and the University of Tokyo. While Fouchier has used a technique called passaging—the H5N1 virus was transmitted and mutated through several generations of ferrets—Kawaoka had directly reassorted the genetic sequence of the H5N1 with nucleotides from the H1N1 and then confirmed its transmissibility on ferrets.[9] It can be argued that Fouchier's research was more "natural," considering he used animals as experimental tools, while Kawaoka's was more "artificial," because he intervened directly at the molecular level; but both experiments tried to simulate what could happen outside the lab through indirect and direct techniques of intervention in living material.[10] It must also be added that in both Fouchier's and Kawaoka's research, ferrets did not die of the mutant H5N1 virus; their results showed the airborne transmissibility of this virus, but not its lethality, which had to be investigated through other researches.

I was intrigued by the fact that virologists expressed these complex relations between nature and the lab through what appeared as collective jokes or Freudian slips. It seemed that the molecular language of virus mutations allowed scientists to map relations between animals and humans that remain unstable in their material and economic dimensions. In a short but dense conversation I had with him, Fouchier explained to me that the price of lab ferrets had increased because they needed to be seronegative to flu viruses, following costly measures of biosecurity, and because they were in competition on the husbandry market with the ferrets raised for fur, particularly in Scandinavia, China, and the United States. While ferrets had long been used in Europe for hunting rabbits and rodents, in recent decades they have been raised predominantly for fur and for lab experimentation. The ferret is the experimental model for influenza research, given that it is the only mammal that sneezes when infected by the virus as humans do.[11] A new page of this fascinating story opened with Fouchier's research, inventing assemblages of cages with ferrets and pigs to show the airborne transmission of flu viruses between mammals.

What had started as a slip of the tongue in face of an ethical requirement thus revealed a complex view of relations between humans and animals. Anthropology has analyzed how statements with double meanings can be understood as conflicting different worldviews. If I cannot go further than this statement as an anecdote, for lack of a proper ethnography of Fouchier's lab, I want to start from the double meaning of "red herring" to enter the ethical debates about the political stakes of Fouchier's research.

Fouchier's work has received a lot of attention in the public debate because it strikingly casts the risks of biological research at the global level. The question, for many observers of the controversy, was not whether Fouchier had successfully replicated in the lab something that could happen within nature, but whether he had been cautious enough in the circulation of biological information.[12] It came to be problematized under the new norm of *biosecurity*. This term covered a conflation of different practices, from biosafety practices in the laboratory controlling the potential escape of pathogens to global monitoring of emerging pathogens, by passing measures to protect national territories against invasive species or contaminated food products.[13]

By the end of 2011, Fouchier's and Kawaoka's research had been examined by the National Scientific Advisory Board for Biosecurity (NSABB). This commission was in charge of checking that the U.S. National Institutes of Health, which had partially funded Fouchier and Kawaoka, would not produce "dual use research." Their members recommended that Fouchier's and Kawaoka's publications should not include methodological details that would enable the reproduction of the experiment by "those who would seek to do harm." A controversy followed in scientific journals on the legitimacy of censorship in biological research, opposing the benefits for public health and the risks for national security of the knowledge on virus mutations. After a meeting at the World Health Organization (WHO) in February 2012, however, the NSABB accepted the full publication of both articles in *Nature* and *Science* and only asked for a six-month delay.

The controversy took a second spin when Fouchier and Kawaoka proposed a one-year moratorium on research in "gain-of-function."[14] They referred to the Asilomar conference in 1975, during which bacteriologists had reflected collectively on the risks of releasing biotechnological mutations in nature. Marc Lipsitch and Alison Galvani, professors of epidemiology at the schools of public health of Harvard and Yale, published an article in which they assessed the risk of research on "potential pandemic pathogens" (PPP). Recalling the dramatic episodes of the release of H1N1 from Soviet labs in 1977, of smallpox from a British lab in 1978, and of SARS from a Taiwanese lab in 2004, they showed that the proportion of accidental release in biosafety level 3 laboratories in the United States between 2004 and 2010 was 0.2 percent, from which they induced that mutant pathogens cultivated in ten labs during ten years had a 20 percent chance of accidental release.[15] Simon Wain-Hobson, an HIV expert at the Pasteur Institute in Paris, asked in a letter to *Nature* on March 27, 2013, "Is it appropriate for civilian scientists to make microbes more dangerous? Is creating a novel human virus antisocial? Was there a failure of duty on the part of funders and regulators? What is the ethical position of such work?"[16]

These positions in the biosecurity debate reflect the techniques of government I have distinguished in the previous chapter. Lipsitch and Galvani practiced prevention referring to statistics of previous cases and asked for a normative regulation of research on pandemic pathogens. Wain-Hobson used a precautionary principle to warn against the fact that the work on mutant H5N1 was "anti-social" and asked for the ethical position of its authors. Fouchier and Kawaoka justified their research by the need to prepare public health authorities for the emergence of a transmissible and lethal avian flu virus from China. While epidemiologists calculated the risk of a pandemic pathogen in the human population, virologists simulated the emergence of this pathogen in the animal reservoir. The language of risk was confusedly adopted by both groups in the precautionary space opened by the media, but they used different modes of thinking to mitigate uncertainties in cross-species transmission of pathogens.[17]

In 2013, Fouchier and Kawaoka wrote in *Nature* to justify their experiments on the mutant H5N1: "Classical epidemiological tracking does not give public health authorities the time they need to mount an effective response to mitigate the effects of a pandemic virus. To provide information that can assist surveillance activities—thus enabling appropriate public health preparations to be initiated before a pandemic—experiments that may result in gain-of-function are critical."[18] "Classical epidemiologists" Lipsitch and Galvani responded to this argument: "Current surveillance is likely inadequate to detect an emerging pandemic strain before it is too late, regardless of any warnings that Potentially Pandemic Pathogen experimentation might generate about potentially worrisome mutations."[19] Here we see a conflict between two definitions of surveillance. Surveillance can mean collecting data on humans to adapt public health policy or collecting data on animals to send early warning signals. In the first view, a particular pathogen is insufficient, since large series of data are necessary, and it can be dangerous; in the second view, creating a pathogen opens new data in a virtual space of monitoring, since it provides tools to follow each nucleotide of an emerging virus and compare it to a target pathogen. These are different views of timeliness—when is it too early to launch the alert?—but also of the extension of surveillance: how to equip animals so that they send adequate signals to humans? Other questions come out of this ambiguity. When does surveillance fail? When does it produce data that do not convert into meaningful signals? And how do actors of public health react to false alarms? How can virologists be lured by wrong targets?[20]

Another key player in the debate was Peter Palese, professor of microbiology at Mount Sinai School of Medicine in New York. Palese was a member of the team, led by Jeffery Taubenberger and Terrence Tumpey, that reconstituted the 1918 "Spanish Flu" H1N1 virus from the frozen corpses of American soldiers.[21] He had adapted the technique of "reverse genetics" to influenza viruses in such a way that this viral material could be injected into chicken eggs and replicated.[22] His research had been exposed to biosecurity concerns in 2005 when it was published in *Science*, and the NSABB had already been involved. This explains why Palese strongly supported Fouchier and Kawaoka in January 2012: "Publishing those experiments without the details is akin to censorship," he wrote in *Nature*, "and counter to science, progress and public health."[23] Palese had been criticizing virologists who warned about the pandemic potential of H5N1 for the last ten years, but he was in favor of replicating experiments on flu viruses, even if he did not share their concern for the lethal effects of cross-species transmission. Palese was attentive to the "semiotics of security" that produced viral mutations in the lab because it allowed him to understand the mechanisms of past pandemics,[24] but he didn't believe that these mutations in the lab could reflect what will happen in future pandemics. Palese believed that the laboratory could replicate nature but not that it could anticipate its mutations—a view that I would call precautionary and naturalistic. Palese was more a "microbe farmer" than a "virus hunter":[25] after his initial success in the reconstruction of the H1N1 virus from frozen human bodies, he was not interested by the relations between humans and animals as they are expressed by viruses in the lab and in the wild.

By contrast, Fouchier and his colleagues designed techniques to imagine and capture future mutations of flu viruses among species. A team of virologists headed by Derek Smith at Cambridge University, also a member of the Antigone consortium, proposed to simulate the risks of emergence of the mutant H5N1 in nature. They didn't want only to calculate and modelize those risks so as to balance "risks in the lab" with "risks in nature," but they designed a technique to visualize the effects of the emergence of a new strain. When I visited him in Cambridge, Smith showed me the software he had built to anticipate emerging viruses. Among the old files and specimens of the zoology department, members of his team sat in front of computers and downloaded sequences of avian influenza viruses to check if they had one or several of the five nucleotides that give the mutant H5N1 a pandemic potential. According to Smith, Fouchier and Kawaoka had provided the world of risk management a virtual target that would select among all potential pandemic viruses. "We are not limited by computational capacities but by the number of samples we have and the number of strains we are targeting," he told me.

Smith thus reflected a position that is dominant in the field of flu research: it is necessary to accumulate the largest collection of flu samples to visualize the paths it might take in the future. This position had been invented by Robert Webster, considered as "the pope of influenza research." Webster chairs the Department of Infectious Diseases of St. Jude Hospital in Memphis, Tennessee. For the last fifty years, he has gathered the largest bank of viral strains, with more than 12,000 samples, to study the molecular evolution of flu viruses and warn public health authorities about their pandemic potential. Although he shares the same ontology of viral mutations as Webster, Palese has appeared as a "counter-prophet" in this world of prophetic truth-claims about pandemic influenza. What Webster and his followers describe as "jumps between species" made by viruses themselves are interpreted by Palese as "leaps of faith" from the rational observation of viral mutations in the lab to the apparently irrational call for pandemic preparedness in the media.[26]

Webster and his followers share a common view of nature that separates them from their critics. It can be summarized in the often-quoted sentence "Nature is the greatest bioterror threat."[27] This sentence means that pandemics would not be initiated by the intentional use of already known pathogens, such as anthrax or smallpox, but by the emergence of new pathogens in nature, such as Ebola in Africa or avian influenza in Asia. It may sound paradoxical, since it seems to attribute intentions to microbes—and biologists play a lot on these paradoxes and metaphors when they say that viruses "hijack" the cells by using their metabolism to replicate or when they describe the pathways of the organisms from the perspective of the viruses that want to invade them. But it is rather embedded in a Darwinian view of the random process of viruses: if they find a niche among humans, as influenza viruses do when they enter industrial poultry farms where vaccines and antibiotics have created a strong evolutionary pressure, silent mutations can produce catastrophic effects. This is what Frank Macfarlane Burnet, Robert Webster's mentor, called "a natural history of infectious diseases."[28] While biosecurity interventions use the rhetoric of states of emergency "whereby the world's disorder, whether natural or human in origin, become equated," thus producing "a naturalization or depoliticization of war,"[29] microbiologists rather introduce war into nature when thinking about the evolutionary process of viral mutation. The metaphors of war applied to virology are valid only in the precautionary space where competing versions of surveillance are mixed.

The Fouchier controversy should be understood from within this vision of nature. Is the mutant H5N1 a good target to track future viral mutations, or is it a lure that distracts virologists? When presenting his research on the mutant H5N1 at the Antigone meeting in Paris, Derek Smith said it was a "quasi-species" to show that its five nucleotides made it different from other H5N1 viruses.[30] Other biologists in the room challenged this statement by arguing that what is present in nature is a "viral mixture" in which all kinds of mutations are present. Smith replied that the evolutionary fitness of the mutant H5N1 within a population of viruses made it specific, arguing that his team had modelized what happens when the mutant H5N1 appears in the cough from a mother to her child and competes with other viruses to become transmissible.[31] The description of a virus as a quasi-species doesn't assume it is an enemy endowed with intentions, but enlarges the possibilities of tracking a target in a space of simulation.

JUMPING ACROSS SPECIES BORDERS

The Fouchier controversy reveals the tensions of biological research about emerging viruses when it is exposed to the public. It stresses the difficulties of communicating with public health officials about preparedness for disease outbreaks. These difficulties were often bypassed through the romantic image of the *virus hunter*. This term was coined in 1960 by Greer Williams after Paul de Kruif's 1926 bestseller, *Microbe Hunters*.[32] Born in the Netherlands, Paul de Kruif played a leading role in the Rockefeller Institute in the United States, and his popular books inspired the vocation of several generations of microbiologists.[33] He portrays the founding figures of Leeuwenhoek, Spallanzani, Pasteur, and Koch as solitary and obsessive men, skilled at inventing new devices to track invisible small beings. Greer Williams recalls that viruses, first defined as nonfilterable pathogenic agents, were made visible only in 1935 on the tobacco leaf[34] and that their molecular structure—a piece of genetic information in a capsid—was modelized only after the World War II. The expression *virus hunters* took a new meaning in the 1990s with a series of books on emerging viruses. Joseph McCormick and Susan Fisher Hoch told the story of their quest for the viruses that were causing hemorrhagic fever in Africa, such as Lassa or Ebola; Robert Gallo described the race between his team and that of Luc Montagnier at the Pasteur Institute to identify the retrovirus causing HIV/AIDS.[35] The power of these narratives comes from the idea that viruses are there "in the wild" waiting to be collected and replicated in the laboratory. What has changed with emerging viruses is the possibility of sequencing them and comparing their forms along evolutionary lines.

One of the main thinkers of the paradigm of "emerging infectious diseases," Joshua Lederberg, who received the Nobel Prize in 1956 for his research on microbe genetics, thus remarked that "during the early acme of microbe hunting, from 1880 to 1940, microbes were all but ignored by mainstream biologists."[36] With the discovery of nucleic acid, microbes became models to test the hypotheses of evolutionary biology. When new viruses emerged in the 1970s, Lederberg acknowledged that it was time to abandon the old metaphor of war between microbes and humans to promote "a more ecologically informed metaphor, which includes the germ-eye's view of infection."[37]

The most popular supporter of the term *virus hunter* is probably—for an English-speaking public—Nathan Wolfe. After starting a career as a zoologist at Oxford University, he claimed, "The study of viruses provides a scientist with the opportunity to discover new species and catalog them in a way reminiscent of the world of the nineteenth-century naturalist."[38] In a series of articles with the equally famous ornithologist and geographer Jared Diamond, Wolfe proposed to modelize the emergence of pandemic viruses from animal reservoirs around the globe.[39] He then set up a private company, Global Viral Forecasting Initiative, whose goal was "to hunt down these events, the first moments at the birth of a new pandemic, and understand them to stop them before they reach a global level."[40] His books and Ted Talks have met with important public success.

Defining himself as a virus hunter, Wolfe is fascinated by the idea that viruses are particularly transmitted by hunting practices, either those of humans who hunt monkeys for bushmeat or those of monkeys themselves, since, he argues, "hunting emerged in our joint ancestors before we split with them."[41] Wolfe describes hunting as an intimate relation between two living beings in which one spends a lot of energy to reproduce at the expense of the other. "From the perspective of a microbe, hunting and butchering represent the ultimate intimacy, the connection between one species and all the various tissues of another."[42] Under the same inspiration, Wolfe portrays viruses themselves as hunters who monitor their environment to find the best pathway through which they will attack their prey.[43] We are then caught, reading Wolfe, into a vertiginous cascade of identifications: virologists are hunters following bushmeat hunters following hunting monkeys following hunting viruses. There are two ways to stop this cascade of identifications and ask about its meaning. One is to actually follow bushmeat hunters to look at their relations of distance and proximity with animals. This is not what I am doing, and it raises ethnographic challenges given the regulations on bushmeat hunting and their different modes of contact with animals.[44] The other is to compare what virus hunters say about animals with what some anthropologists have described about hunting in general as a mode of relation to the environment. This is what I call "taking virus hunters seriously" to understand their ontological disagreements with epidemiologists on the public-health impact of monitoring animal diseases.[45]

What does it mean to look at the relations of proximity and distance between humans and animals from the perspective of microbes that emerge in these relations? What kinds of signs are produced by humans, animals, and viruses if they hunt each other, and how do these signs act in specific contexts? These questions find answers in an anthropology of shamanism, defined as a technique to regulate relations between humans and animals. While this notion was coined in the eighteenth century after a Siberian term and popularized after the Second World War by Mircea Eliade to describe an archaic religious state characterized by initiatory trance, it has recently been used by some anthropologists to describe relations between humans and animals in Amazonian and Siberian societies. For hunters in these societies, the relation to animals is particularly unstable: they need to see the environment through the perspective of the animals they track in order to catch them; but they should not see themselves through this perspective, otherwise they would consider themselves as potential prey. By contrast with pastoral societies, in which humans are supposed to be above the animals they domesticate, shamanism is a technique to manage uncertainties in societies where relations between humans and animals are reversible.[46]

Recent ethnographies of shamanism in Amazonia and Siberia, rather than engage the speculative paradoxes of shamanism, have showed its effects on the language used to describe relations between humans and nonhumans. Eduardo Kohn has described the different levels of signs that constitute reality for the Runa of Ecuador, whose moral responsibility is to give signs an adequate form.[47] Morten Pedersen has coined the term *not-quite shamans* to describe "inexperienced hunters who offend the spirit masters of the game, or the targets of mischievous jokes and malicious slander: their souls are all too easily lured away."[48] Charles Stépanoff suggests that because shamans are conceived as belonging to a different species from humans, their capacities to perceive invisible beings are attributed to an inborn abnormality. True shamans can thus be distinguished from false shamans by their ritual body, built out of accessories such as drums or weapons or animal parts, that magnifies this inborn specificity and meets the challenges of encounters with supernatural beings.[49]

Inspired by these ethnographies, I want to show that the double meanings of statements produced by virologists reveals a conflation between "shamanistic" techniques of preparedness and "pastoralist" techniques of prevention. I take the controversy between Fouchier and his critics on the responsibility to spread information about potentially pandemic pathogens as a starting point to describe the modes of equipment they have designed in the laboratory and "in the wild" to prepare the globe for future pandemics. I thus move beyond the linguistic analysis of arguments in a controversy to analyze the modes of collecting of virus hunters.

HUNTERS AND COLLECTORS IN AUSTRALASIA

While the Fouchier controversy was mostly held in the United States, as it involved agencies such as the NIH and the NSABB, its main actors came from Europe and Asia. While most of the discussions in the U.S. turned around the threat of bioterrorism, which had grown after the fall of the Soviet Union and the episode of "anthrax letters" after 9/11, in Asia they were concerned with the consequences of the SARS crisis, which had revealed the vulnerabilities of Asian societies to emerging viruses and was often presented as an "Asian 9/11."[50] SARS, spreading from bats to humans in the complex ecology of south China, revealed the necessity to map nature in order to understand how it had become "a bioterrorist threat."

The members of the Antigone consortium with whom I worked in Europe had been key actors in the SARS crisis. Albert Osterhaus was the director of the virology department of the Erasmus Medical Centre since 1993, and is described as one of the most famous virus hunters in the world.[51] A veterinarian by training, he has studied the viruses of seals and dolphins in the South Sea with the support of Dutch animal-protection movements.

After the crisis caused by mad cow disease in Europe, he warned in 2001 about future pathogens crossing species barriers.[52] In 2003, he was the first to apply Koch's postulates to the SARS virus, successfully transmitting it to monkeys.[53] In the same year, Holland was facing a new avian influenza virus, H7N7, which killed millions of chickens and one veterinarian. After this striking event, the only human death caused by an avian flu virus in Europe, Osterhaus engaged himself in the global warning against the threats of H5N1. During the H1N1 pandemic in 2009, he designed a computer game that simulated the sale of masks, vaccines, and antivirals, which made him a target for members of the European Parliament, who criticized his links with the pharmaceutical industry. He recently left the direction of the Erasmus Medical Centre to found a private research center in Hamburg, called Viroclinics.

Another key actor of the Antigone consortium was Christian Drosten, head of the Institute of Virology of the University Hospital of Bonn. After he developed the first diagnostic test for SARS on March 2003,[54] he launched an ambitious program for sequencing viruses in bats in Ghana and Brazil. With the support of associations for bat protection in Germany, he has discovered that German bats carry coronaviruses similar to SARS.[55] He has developed collaborations with Australian and Chinese virologists who collect and sequence samples from bats in Southeast Asia to understand how they have become the animal reservoir for many emerging viruses, such as Ebola, SARS, and Hendra and Nipah viruses.[56] Bats are a particularly fascinating model of animal reservoir because of their important species diversity and their complex immune system. It can be hypothesized that after millenniums of multi-species co-evolution in caves and forests, and to meet the challenges of flying for a mammal organism, bats have developed a refined immune response to emerging viruses, in comparison with which human immune systems appear rudimentary.

Germany, Holland, Great Britain, and France have a long history of colonial microbiology linking them to the Asia-Pacific area in the fights against cholera, plague, and malaria at the end of the nineteenth century. In these medical campaigns, tropical diseases are considered obstacles to the expansion of European colonialism because they develop in specific milieus. However, the Antigone project is rather inscribed in a postcolonial temporality whereby Asian territories are considered sentinels for the early detection of global diseases because they are at the heart of current ecological changes. In an influential and precursory paper published just before the SARS crisis, Nicholas King identified the "emerging diseases worldview"

FIGURE 2.1. The WHO expert committee on influenza in Geneva in 1978, headed by Martin Kaplan. Kennedy Shortridge is fifth from the left, Graeme Laver and Robert Webster are third and second from the right. Courtesy of Kennedy Shortridge.

as a "tremendously flexible" view of the world, "allowing a wide variety of actors to adopt it . . . a consistent, self-contained ontology of epidemic disease, . . . equipped with a moral economy and a historical narrative, . . . a universalizing template for understanding the interactions between humans and the microbial world."[57] In contrast with a colonial vision of public health, which imposes knowledge from the center onto the periphery and aims to preserve territories, this new rationality of risk aims to collect and disseminate information in deterritorialized networks. "Surveillance is imagined to be everywhere, at all times, producing data available to everyone: a global clinic."[58]

Robert Webster, "the pope of influenza," starts the narrative of his career at the beginning of the 1960s with the following event, which inaugurated a new set of surveillance practices. He was walking along a beach near Canberra, where he was doing his PhD in microbiology, with his university colleague Graeme Laver, who was working with him on the molecular analysis of influenza viruses. As they saw dead seabirds on the shore, they joked that these birds had died from flu.[59] They consequently launched a massive program of trapping and sampling seabirds in the Great Barrier Reef to survey the patterns of influenza A in birds (by contrast with influenza B and C, which only circulate among humans). Martin Kaplan, a veterinarian who was the head of the WHO zoonoses program in Geneva and who was later replaced by Webster in the 1970s as the head of the WHO Centre on the ecology of influenza, gave them funding for this project, which had opposed the skepticism of local authorities.[60]

Graeme Laver recalls how the head of his department of microbiology reacted to his project. "'Anyway there is no way he is going to be able to catch the birds!,' I wasn't so stupid. I knew the shearwater nested in burrows in the ground and all we had to do was bend over and pitch them up. But I must admit the thought of those beautiful healthy birds on a deserted coral island surrounded by the bluest of blue seas under a scorching sun carrying influenza viruses was almost too bizarre to even contemplate seriously."[61] Webster and Laver imposed this "bizarre" idea that aquatic birds could be the reservoir of influenza, meaning that they carry the virus without being sick and shed it massively via fecal drops. Warning of the future threat of pandemic in 2003, Webster wrote, "A major challenge in controlling influenza is the sheer magnitude of the animal reservoirs."[62] In the obituary he wrote for Laver, Webster praises, "His Australian heritage of adventure in the great outdoors led to studies on influenza in migratory birds."[63] The history of flu research is thus part of what historian Tom Griffiths has described as "a refined hunting and gathering" among Australian naturalists.[64]

My analysis of the Fouchier controversy thus suggests that the work of microbiologists should not be read only in connection with that of public-health officials (epidemiologists, pharmaceutical industries, hospital administrations, experts in risk communication) but also as borrowing practices from those who work with animals (veterinarians, meat traders, farmers, and naturalists). If surveillance and anticipation have become the key words and cornerstones of global health, they are not new techniques oriented toward the future, but a recombination of heterogeneous techniques from the past. The Fouchier and Kawaoka controversy, which started as a public debate about biosecurity between Europe and the United States, appears as a starting point for an investigation on what it means to do surveillance of natural mutations in Asia and Australia.[65] In chapter 3, I will argue that this effort is part of an attempt to collect and compare samples from Asia and that avian influenza research is not only oriented by the ideal of a "global clinic" but also by the idea of a "global museum." Preparedness will thus appear not only as a technique of government producing slips of the tongue and contradictory statements, but also as a way of seeing and classifying the beings of the world.

CHAPTER THREE

global health and the ecologies of conservation

This book describes one of the recent articulations of global health. Such a term encompasses the multiple initiatives to reorganize public health beyond the level of nation-states' populations and to reach individuals in the specific diseases that affect them. Covering a wide assemblage of actors, from biomedical experts to development agencies including philanthropic foundations, it is split between the anticipation and control of emerging infectious diseases, oriented toward security in the North, and the compassion and care for any victim of disease, oriented toward humanitarian action in the South. These "two regimes of global health," Andrew Lakoff argues, seem to be in conflict, but may also work in complementarity. "Humanitarian biomedicine could be seen as offering a philanthropic palliative to nation-states lacking public health infrastructure in exchange for the right of international health organizations to monitor their populations for outbreaks that might threaten wealthy nations."[1] The compassion to victims in the global South and the desire of security in the global North represent the two faces of the global subject: the suffering individual and the vulnerable infrastructure.

This book makes a similar argument about global health when it extends to animal diseases and environmental issues. Under the label "One World, One Health," international organizations (WHO, OIE, FAO) have shared their systems of information and connected their networks of surveillance to cover all aspects of the health of humans, animals, and their environment.[2] This global project met with the needs and requirements of a wide range of actors, from development organizations in agronomy such as Vétérinaires Sans Frontières and GRAIN[3] to environmental networks such as the Wildlife Conservation Society, the World Wildlife Fund, and BirdLife International.[4] "One World, One Health," with its set of techniques for monitoring and surveillance, thus mixes two different rationalities: anticipation of emerging pathogens and conservation of the environment. When global health thus extends to animals and the environment, it conflates two different rationalities of risk. One, oriented toward preparedness, follows the mutations of pathogens between species and requires the work of microbiologists to anticipate the next catastrophe. The other, oriented toward prevention, counts the number of casualties of ongoing disasters with the help of epidemiologists and tries to mitigate these disasters by curing the victims. The first aims at biosecurity—the control of the proliferation of living material circulating on the globe—while the other aims at biodiversity—the inventory and conservation of the various forms of life inhabiting the planet.

What does it change, then, to look at global health from the perspective of conservation rather than compassion? How is global health redefined from the multiple sites dedicated to the conservation of nonhuman forms of life? In this chapter, I propose a genealogy of avian reservoirs in the Asia-Pacific by tracing how conservation practices in Western societies have been shaped in relation with samples coming from the Eastern part of the world. If birdwatchers work in wildlife reserves and microbiologists in scientific laboratories, these modern spaces for the observation and accumulation of natural forms derive from the ancient *locus* of the museum. I will show how the museum, defined as a space for the conservation of living forms, has been transformed in parallel to the hospital, defined as a space of compassion for suffering victims, when prevention has been replaced by preparedness. This chapter will thus propose a genealogy of global health and global art as two domains for an anthropology of the contemporary. While "global health" watches over the mutations of natural reservoirs, "global art" investigates the transformations of cultural heritage. Both anticipate the future through the imagination of the present. If museums, laboratories, and reserves can be described as spaces where artifacts are collected, stored, and displayed, how have they been redefined by techniques of preparedness as a new imperative to anticipate the mutations of pathogens and share the information about them? This parallel genealogy will consider separately three forms of Western knowledge as they are extended to China: virology, ornithology, and anthropology. By questioning historically the practices of collecting and curating involved in these knowledges, I will ask ontological questions: What is a virus? What is a bird? What is the social?

MUSEUMS OF VIROLOGY

Among other forms of storage (from repositories to banks), the museum can be defined as "a way of seeing"—that is, the organization of distinctions between forms on display creating an aesthetic pleasure.[5] Rather, it is a site of competition between ways of seeing and a crucible for shifts in the imagination of the future. The vocation of the museum, since the Greek creation of the term, is to educate the eye so that it sees forms in objects coming from natural hazards or from cultural intentions. As such, museums' educative function had been defined well before the opposition between art and science, which is a product of modernity. But with the advent of capitalism and nation-states, museums have been endowed with a new function: accumulate and conserve artifacts to display the power and wealth of the sovereign. If museums result from hunting practices—collectors have to track forms in their natural environment and trap them in a new cultural setting—they have been reframed within a pastoral framework: objects now appear as a set of singularities with names and numbers that have to be conserved and taken care of. When they entered modernity, museums consequently became sites of application for the new rationality of prevention. Because they displayed natural forms in ways that the eye could enjoy and that the intellect could master, they allowed humans to foresee the future as a series of calculable risks.

The framework of the museum was applied to the research on influenza by the World Health Organization after the Second World War. This research had burgeoned in the interwar period, connecting microbiologists and epidemiologists to understand the mechanisms of the 1918 "Spanish flu" and the cyclical character of flu pandemics.[6] The first isolation of a flu virus in a British laboratory in 1933, by inoculating ferrets with human mucus, was deemed successful because, by contrast with other experimental models, ferrets sneeze as humans, their temperature rises, and their nose runs when they have the flu. Carlo Caduff comments, "The pathological effect became effective and provided evidence because it took the right shape in a framework of visibility that was constituted by the clinic and that relied on the set of symptoms observed by humans."[7] Transferring influenza viruses from the clinic to the lab through animal models allowed microbiologists to compare them and organize their visibility, thus shifting from symptoms of a current disease to signs of future pandemics.

The U.S. Army set up a Commission on Influenza in 1941 headed by Thomas Francis, who had confirmed the British isolation of the influenza virus with the support of the Rockefeller Foundation.[8] Among this team was a young virologist named George Hirst, who had discovered a technique to identify whether a flu virus was present in a human sample. Hirst had exposed a chicken egg containing a living eleven-day embryo with a flu virus and observed that the red blood cells of the embryo were sticking together as if to defend themselves against the virus, a reaction he called "agglutination" and that was later reframed as "hemagglutination."[9] This discovery allowed Hirst to design a test (called an HI assay), which combined the injection of flu viruses with the injection of human serum containing antibodies in a chicken embryo and measured the antigenic differences between viral strains. The experimental model in the lab thus made visible and explained the obstacle faced by clinicians and epidemiologists: influenza viruses were constantly mutating, which made it difficult to design a vaccine for the next strains of influenza based on existing strains. If the difference between human strains was visible in chicken embryos in the lab, it became possible to prevent future pandemics by adapting vaccines to the distance between strains. Flu viruses were consequently classified by the surface protein that commanded the hemagglutination when the virus entered the cell, called H, before another surface protein called neuraminidase (N) was identified in the 1960s as playing a major role in the release of the virus from the cell, and then became a target for antivirals such as Relenza. Hence the names given to influenza viruses: H1N1, H2N2, H3N2, H5N1.

The World Health Organization benefited from these discoveries to proclaim after the war that influenza viruses from all over the world should be compared to adapt vaccines and prevent the next pandemic. In 1948, the Centre for Influenza Research of WHO defined itself as a "museum" for flu virus strains and asked laboratories to send them in a dessicated form. Notice the difference between a repository and a museum: while a repository contents itself with storing and conserving materials, a museum has to classify and inventory this material and make its differences visible by displaying them to the public.

> The dessicated viruses received by the Centre, will, as far as possible, be tested to determine their activity and, where appropriate, passaged to form a larger stock of dessicated material. It will thus constitute a sort of "museum" of dessicated strains of influenza. All laboratories that conduct research on the antigenic relations between strains may request the Centre to be kept informed of available strains and obtain the sending of a particular strain. We will periodically submit the museum strains—or at least those which seem to be important and representative—to passages in eggs or others to maintain stocks and avoid losses.[10]

The technique of egg passaging designed by George Hirst was thus applied by the WHO Centre for Influenza to dessicated strains they received from labs; in return, laboratories from all over the world could compare the strains they received from clinics to reference strains. Just as art museums were being redefined by the International Council of Museums of UNESCO in 1945 as places where works of art were conserved and should circulate through temporary exhibitions, virology laboratories were redefined by WHO as places where flu viruses should safely circulate after major outbreaks before being compared at a central level. Virologists became the managers of a global circulation of viral strains, while a massive industry of chicken eggs, killing billions of chicken embryos, made possible this global production of knowledge, as the invisible working force underlying the visibility of the museum. Carlo Caduff writes about this circulation of flu virus strains between labs: "Since 1933, biomedical scientists and public health experts have invested considerable resources in generating and channeling a seamless flow of viral strains not only across species but also across countries, institutions and disciplines. It was this controlled flow of biological matter that allowed influenza research to become independent of the seasonal occurrence of epidemic disease."[11]

A major shift occurred at the end of the 1970s in influenza research, revealing the flaws of this dream of a universal museum of flu strains set up by WHO. The emergence of an H1N1 "swine flu" virus in the U.S.—possibly from a "Spanish flu" virus released from a Soviet lab—triggered a massive campaign of vaccination after the death of a soldier in Fort Dix from this new virus in 1976. This campaign, known as "Swine Flu Fiasco," was interrupted when 10 percent of the population had received the vaccine and more than 500 cases of Guillain-Barré syndromes were declared.[12] This event showed the difficulties of preventing the next pandemic with vaccines based on the differences between strains. It also revealed that circulating flu viruses between labs was not neutral, since virus strains, when they are alive, can escape from the labs and replicate in the population to cause new pandemics, which led to increasing concerns about bioterrorism after the end of the Cold War.

Another parallel event changed the strategy of the museum of viral strains. In 1976, WHO opened an Expert Committee on the Ecology of Infectious Diseases under the leadership of its chief veterinary officer, Martin Kaplan. The role of this committee was to anticipate the emergence of new pathogens from animal reservoirs where they mutate. Among the key members of this committee were Graeme Laver and Robert Webster, who had set up a bank of influenza virus strains by collecting bird feces all over the world with the support of Martin Kaplan (see chapter 1), but also Kennedy Shortridge, also trained in Australia under the mentorship of Frank Macfarlane Burnet, who had just opened the Department of Microbiology at the Faculty of Medicine of the University of Hong Kong. As we will see in chapter 4, Shortridge's idea was that since China was not part of WHO, and since major flu pandemics originated from south China that he considered an "influenza epicenter," there was very little knowledge on the flu strains circulating in China. Hence Hong Kong should become a sentinel post where flu strains could be collected, compared, and sent to WHO to contribute to the global museum.

These two events can be described as operating a shift from prevention to preparedness in the strategy of WHO as a museum of flu strains. The first event showed that the circulation and comparison of flu strains raise safety issues, since viruses have a life of their own that cannot be reduced to the distances between antigenic forms: they mutate, reassort, escape, trigger unexpected immune responses. The second event revealed that a large part of the global circulation of flu viruses remained unknown, because nation-states such as China refused to provide them or simply because they didn't have the means to collect them. When there are gaps in a collection, strategies must be found to cover these gaps.[13] The invention of stockpiling vaccines was a response to the first event, and the implementation of sentinel devices was a response to the second. If the next flu pandemic could not be prevented based on the comparison between existing flu strains, it became necessary to prepare for it by capturing early warning signals to mitigate the consequences of the outbreak.

Simulation of pandemics thus became the major technique of influenza management at WHO. This became possible through GenBank, which has allowed virologists to compare in real time a new strain to existing strains based on their genetic sequences and to imagine through bioinformatics software where this new virus comes from and how it could evolve. In an article entitled, "The Experimenter's Museum," Bruno Strasser has showed how GenBank has changed what he calls "the moral economies of biomedicine." Rather than replacing eighteenth-century natural history, Strasser argues, experimental biology has rearticulated its method of building collections in a center reflecting the diversity of the world: "Those who created the early modern cabinets of curiosity, the royal gardens of the seventeenth and eighteenth centuries, and the great zoological museums of the nineteenth and twentieth centuries all faced the challenges of bringing to a central location specimens that were often dispersed all over the world, securing the participation of individual naturalists, and negotiating the status of the specimens in their collections."[14]

While natural history collections displayed the power and wealth of those who accumulated them, Strasser shows that contemporary databases have reorganized this modern dream of the collection through a new moral economy of science, around the participation of all citizens to the production of science. The organizer of the database should not be the owner of the collection, but the manager of a collective effort in which every individual initiative is valorized. Strasser thus contrasts two genetic databanks that were in competition in the U.S.: that of Margaret Dayhoff at the National Biomedical Research Foundation in Washington, D.C, starting the publication in 1965 of an *Atlas of Protein Sequence and Structure*, and that of Walter Goad at the Los Alamos Scientific Laboratory in New Mexico, using radiation to study nucleic acid sequences of proteins: "Neither Dayhoff nor Goad had ever sequenced a protein or a piece of DNA; they relied on others to accomplish that. Goad tried to acquire sequences in bulk from other collectors, whereas Dayhoff obtained them by searching the literature and through daily interactions with those who had determined sequences in their laboratories."[15]

While Goad was ready to share his small collection of protein sequences, Dayhoff considered herself the owner of her bigger collection, to which she could refuse access. One of her interlocutors among biologists reproached her for constituting her own "private hunting ground," a criticism often made by collectors when they describe themselves as hunter-gatherers.

Consequently, in 1982, the National Institutes of Health funded Goad's database and not Dayhoff's, which led to the development of Genbank at Los Alamos before it moved to the National Center for Biotechnology Information in 1992. Strasser argues that Goad successfully adapted the practices of museum collectors to the new moral economy of experimental science and that Dayhoff kept eighteenth-century ideas about the property of collections.[16] I would rephrase it by saying that Goad found a stable compromise between the pastoralist needs of institutions such as the NIH and WHO and the practices of virus hunters, whereas Dayhoff proposed a precautionary view of science that was unsatisfying for both parties. For biologists, the value of their findings comes from their success in tracing the movements of a living entity through genetic sequencing, while for public health authorities it comes from the idea that this sequence takes part in a databank in which it was missing.

MUSEUMS OF ORNITHOLOGY

To support this claim, I now turn to the practices of birdwatchers. The argument of this book is that virologists identify themselves with viruses when they track them in animal reservoirs, just as ornithologists identify with birds when they follow their movements in natural reserves. Both scientific practices come from modern natural history, and yet they have been profoundly transformed by the contemporary use of databases, which allow them to follow their targets virtually.

Frank Macfarlane Burnet, the founder of the Australian school of microbiology, began his career as a passionate zoologist, particularly collecting beetles in rural Victoria. After successfully transmitting influenza virus in chicken eggs for the first time in 1936—thus building the path that would lead Hirst to his identification technique years later—he promoted the collection of microbiological samples from birds.[17] He opens his *Natural History of Infectious Diseases* by relating microbiology to nature observation in England and other industrial countries.

> Since the eighteenth century, there have always been educated men of some leisure with a natural interest in the activities of animals and plants. Many of these amateur naturalists, from Izrael Walton and Gilbert White onward, have written about the way animals make a living. The habits of birds in feeding, courting and nesting have attracted the interest of many. Others have spent years unraveling the life history of insects. In more recent years,

the essentially amateur type of observer has been supplemented by the professional biologist, whose more systematic investigations in the field once known as nature study have raised its dignity to the science of ecology.[18]

According to Burnet, the observation of nature as a leisure was replaced by a more systematic study of ecology in the laboratory where living beings were rigorously named and classified. Yet Burnet fails to describe the parallel changes in virology and ornithology to which his own findings have contributed. How is birdwatching transformed when it is practiced not in a pristine nature but in an ecology of infectious diseases?

The genealogy of birdwatching practices in Europe shows the role of museums in the emergence of a sensibility to threats affecting birds. Stefan Bargheer recalls that parallel to the rediscovery of the English countryside by naturalists often inspired by theology, the eighteenth-century passion for birds came from the practices of colonial collecting. Joseph Banks, a botanist who later succeeded Isaac Newton as the head of the Royal Society, took part in the first travel of James Cook in the Pacific Ocean between 1768 and 1771 and brought back 500 bird specimens and 32 drawings, which he sent to the British Museum.[19] Bargheer notes that this aesthetic of the museum contrasts British ornithology with its German counterpart as national constructions of nature. In the United Kingdom, birdwatching came from the pleasure to see birds in nature as if they were ordered as a museum in a "local patch"; in Germany, it came from the needs of managing nature as a household or *Heimat*. While British birdwatchers opened the wings of birds to see what they looked like, German ornithologists opened the bellies of birds to see what they ate. In nineteenth-century Germany, ornithologists classified birds between useful and harmful (or pest-eaters and pest-carriers), whereas British birdwatchers made a major distinction between rare and common birds (or exotic and local). While in the UK the value of bird specimens depended on how they completed a collection, in Germany it depended on how useful these birds were for the habitat.

These national constructions should not be taken as cultural frameworks, but rather as results of compositions between hunting practices and pastoral techniques of power that have come to hybridize the modern practice of birdwatching, as revealed by a short comparison between ornithology museums in the UK, Germany, France, and the United States. In France, the most prolific writer about birds in the eighteenth century, and a major source for the classifications set up by Buffon and Cuvier when they organized the Ménagerie of the Jardin des Plantes, was in charge of the royal hunting grounds of Versailles and Marly. Charles-Georges Le Roy is famous for criticizing philosophers who discussed the souls of animals through an empirical argument. "Only hunters can appreciate the intelligence of animals," he wrote in the 1760s. "Knowing them well requires living in society with them."[20] The tradition of hunting for the sovereign as a way to connect game and the territory may explain why in France the Ligue de Protection des Oiseaux has only 35,000 members, while the Fédération Nationale des Chasseurs has 1.3 million members. In the UK, the Royal Society for the Protection of Birds ranks 1 million members (with 800,000 licensed hunters), while the *Naturschutzbund* in Germany boasts 420,000 members (and 300,000 licensed hunters), and the National Audubon Society in the U.S. has 550,000 members (and 35,000 licensed hunters).[21]

These figures can be contrasted with those of ornithology collections: while the British Museum conserves 750,000 specimens of 8,000 bird species (mostly coming from the travels of Cook, Gould, Darwin, and Wallace), the bird collection of the Smithsonian has 640,000 specimens of 8,500 bird species (mostly coming from the expeditions organized by bird curators such as Ridgway and Chapman), the Berlin Naturkünde Museum conserves 200,000 specimens of 5,000 bird species (with a systematic organization of bird collecting by Lichtenstein, Cabanis, and Reichenow), and the Museum National d'Histoire Naturelle in Paris conserves 130,000 specimens of 2,500 bird species (mostly coming from the travels of Baudin, Freycinet, Dupeyrret, Dumont d'Urville, and Bougainville). France is also the only country where a museum dedicated to "hunting and nature conservation," supported by a private foundation of nature lovers and game hunters, is located in the middle of a capital city (Paris's fourth arrondissement), and displays natural specimens with contemporary art to reflect on the ambivalence of killing animals.

The founder of British ornithology, John Latham, transformed hunting practices of bird collectors into preventive rationality at the museum, a hybrid form that was later replicated globally. After studying anatomy with William Hunter, he became a member of the Royal Society in 1775 and published about birds in the society's *Transactions*. He consulted the bird collections sent by Joseph Banks at the British Museum as well as the cabinet of curiosities of Ashton Lever, which also contained specimens from the Cook voyages. He published *A General Synopsis of Birds* between 1781 and 1785, applying Linnaeus's systematic classifications as well as the division between land and water birds proposed by John Ray.[22] As the specimens from Cook's travels rapidly turned into decay, particularly in Lever's cabinet, the handbook published by Latham secured the knowledge about bird diversity. While the cabinet of curiosities displayed birds for their extraordinary aspects without much care for their materiality, ornithology museums defined the rarity of a bird species in relation to the whole collection and conserved its materiality accordingly.

For ornithology museums, the vulnerability of bird materialities (not only the fragility of their skeletons and the organicity of their skins but also the colors of their feathers) was a major problem, faced by the new profession of taxidermy.[23] Since museums were conceived as repositories of knowledge about bird diversity, this rapid decaying reminded ornithologists of the ambivalence of killing birds. American ornithologists were particularly sensitive to this ambivalence because they were first animated by a feeling of abundance of wildlife in the territory of their young nation before discovering that many bird species became extinct after massive hunting campaigns. The first curator in ornithology at the Smithsonian Institution, Robert Ridgway, was passionate about reading books as well as about hunting birds. His biographer, Daniel Lewis, writes that his "study of birds (was) largely initiated by sighting carefully down the barrel of a gun."[24] John Muir described him as having "wonderful bird eyes, all the birds in America in them."[25] By contrast with Audubon, who was mostly concerned with observing and drawing the birds of North America in their habitat, Ridgway was obsessed with classifying, conserving, and exchanging bird specimens in the museum.

The dream of a complete classification of living beings following Linnaeus's and Darwin's principles is expressed in this definition of science by the *American Naturalist* in 1882: "a rationally established system of facts and ideas, which, over a given range of objects, confers certainty, assurance, probability, or even a doubt that knows why it doubts."[26] The museum then became a place to foresee the future evolution of life by considering series of past cases. "Museums' long runs of series of the same birds, . . . allowed scientists to identify subspecies, catching birds in the act of evolving,"[27] Lewis writes. In this quote, the word *catching* reminds the reader that Darwin's evolutionary theory is inscribed in the practices of hunters and collectors. George Good, who was assistant director of the Natural History Museum of the Smithsonian when Ridgway was curator in charge of ornithology, described this tension in the Smithsonian's 1896 *Annual Report*: "The people's museum should be much more than a house full of specimens in house cases. It should be a house full of ideas, arranged with the strictest attention to the system."[28] The evolutionary theory was a criterion to give value to the materiality of bird samples: it allowed curators to hierarchize their stocks of specimens following lines of supposed descent and thus to select among all the specimens available those who had more value than others. Small birds were consequently preferred to big birds by curators, not only because they were less difficult to store, but also because they often filled the gaps in collections and were easier to exchange with private collectors who didn't valorize them.[29]

Yet the materiality of birds' bodies returned in this system of ideas through the challenges of their decay. Lewis notes that naturalists of the museum constantly expressed depression in the summer months, which he explains by the fact that arsenic was used in this month to prevent rotting of the skins and insect infestation.[30] A new preventative against insects and skin deterioration, Maynard's Dermal Preservative, became popular in ornithology museums in the 1880s because it also protected the hands of curators from arsenic poisoning.[31] Curators who took care of bird specimens through the ideal hierarchy of evolutionary systems thus suffered from the same troubles that affected bird skins and could take the perspectives of birds on their own decay.

The tension between a hunting approach to bird collecting and a pastoral approach to bird curating became more vivid in the 1930s as the awareness grew that birds were becoming extinct in their natural habitats. In 1934, the Marquess of Tavistock wrote a letter to the editor of the ornithological journal *The Auk* to complain about "the ruthless and excessive destruction of rare birds by the American Whitney Expedition" in the Pacific Islands.[32] Frank Chapman, curator of the bird department at the American Museum of Natural History, the institution that had sponsored the expedition, responded to deny that the collectors had posed a serious threat to any birds. Chapman strongly believed that bird specimens were more secured in a museum than in the wild because for him they found their place and meaning in an evolutionary framework. This view is clearly expressed by Edward Nicholson, who organized the Oxford Bird Survey in 1926:

> A bird-lover . . . looks at birds as living beings and watches their movements, habits, and signs of life; so that for him only the living bird is a bird at all, after death it becomes merely a corpse. But the collector is always looking at even the living creature in terms of the glass case—he admires a fine specimen far above a fine singer, and to him the live bird is always a chrysalis ready to be transformed into the perfect *imago* which only a spell at the

taxidermist's can make it. . . . When he shoots a bird he is not conscious of having destroyed anything, but only of having secured it.[33]

The most striking case of hybridization between cynegetic and pastoral techniques of power in ornithology may be found in the character of Günther Niethammer, who was an SS officer in the concentration camp of Auschwitz and published in 1941 his "Observations on the Bird-Life in Auschwitz" in the *Annals of the Natural History Museum* at Vienna. Niethammer had been in charge of editing the *Handbook of German Bird-Life* at the Berlin Museum between 1937 and 1942 and had been hired in Auschwitz between October 1940 and October 1941 to shoot at wildlife around the camp. Niethammer presented himself as "a kind of concentration camp SS game-keeper (*Jägermeister*),"[34] thus revealing how the pastoral logic of extermination in Germany relied on hunting techniques.[35]

The global politics of ornithology after the Second World War was in many ways a reaction to the failures of national projects of bird conservation. Under the supervision of the International Union for the Conservation of Nature (IUCN), lists of endangered species were written to rule the exchanges of natural specimens between natural history museums and wildlife reserves. While it seems that wildlife reserves replaced natural history museums as sites to practice ornithology, it may be argued that these reserves were managed as museums—that is, as sites where the diversity of living forms should be monitored. In 1946, Peter Scott opened the Wildfowls and Wetland Trust in Slimbridge to shelter all wildfowl species of the world in Britain. He founded the World Wildlife Fund in 1961 to provide money to conservation projects, and particularly to the IUCN, and set up the red list of endangered species. Projects such as the reintroduction of cranes in Britain involved several countries, including Germany and France.[36] Birds were not valued any more as specimens filling gaps in a collection, but as "indicator species" in a trend to extinction.[37] This notion was introduced in 1973 by Norman Moore, head of the Nature Conservancy's research unit on pesticides, who wrote, "While wildlife becomes increasingly valuable for economic, scientific and aesthetic reasons, it is also developing a new value; technological change today is so extensive and rapid that we require pointers to show us what we are doing and to help us to predict. Increasingly, it is realized that wild plants and animals have yet another role to play in the modern world—that of biological indicators of unforeseen problems."[38]

The notion of indicator species brings hunting techniques of communicating with animals into a pastoral space where "wildlife" is managed as a flock or a population. Hunting techniques have never disappeared in the transformations of ornithology from natural museums to wildlife reserves because birds have always been hunted and monitored at the same time through different combinations of power. Museums defined birds as endangered beings before awareness of bird extinction was expressed by naturalists in wildlife reserves, and these two spaces worked for the preservation of vulnerable species. Ornithology museums, rather than enclosing birds as cultural artifacts removed from their natural habitats, have been the environments in which modern humans became aware of birds' vulnerabilities, which explains why wild bird reserves are managed as museums. In these reserves, birds are observed, classified, cured, and monitored by naturalists as the artifacts of a museum are conserved by curators. Ornithology museums and wildlife reserves have thus gradually shifted from a paradigm of prevention, through the theory of evolution that made series of bird species meaningful, to a paradigm of preparedness, through scenarios of extinction in which the fate of humans and nonhumans is imagined through indicator species. This shift has been clearly expressed by Stephen Moss, one of the most popular birdwatchers and writers in Britain:

> I often think an interesting analogy is with art. Why would it matter if all Raphael's paintings were destroyed? Suppose they have all gathered at one exhibition and there was a fire and they're all destroyed, why would it matter? Because people would be saying: "what a terrible disaster." But there are still photographs of them, there are nice paintings, but so what? But of course it would matter and we would finally inquire to explain why it would matter. I think we would all feel that it was a tragedy for human culture. And I think it's like that when birds become extinct.[39]

By comparing the extinction of a bird species with the destruction of an artist's work, Moss makes an analogy between a natural habitat and a cultural heritage; but he also shows that managers of these sites should prepare for these events—that is, imagine what the world would look like without these valuable beings.

The concept of avian reservoirs thus appears in the 1960s at the crossroads of parallel developments in ornithology and virology. If viruses and birds have been cultivated and classified in laboratories and natural reserves as sheep in a flock, but also monitored and followed as if they evolved in the wild, the idea that new viruses emerge from wild birds in China may be another step in the history of global museums. The encounter between microbiologists and birdwatchers in Hong Kong at the beginning of the twenty-first century is thus inscribed in a longer genealogy of Western collecting practices. The description of China as an epicenter of influenza pandemics may turn out to be one of the drivers to record the proliferation of natural species in China before they become extinct.

Still, there is a difference between this dream of a total archive and preparedness. Preparedness, I have argued, is the return of hunting practices in the methods of collecting that define modern science, as it involves imagining the future from the pathogens carried by birds. The ideal of a universal collection of knowledge has been replaced by the imagination of vulnerabilities in infrastructures. Paula Findlen has showed that the role of collecting in sixteenth-century cabinets was "to establish storehouses to monitor the flow of objects and information. . . . While we perceive the museum of natural history to be alternatively a research laboratory or a place of public education, [sixteenth-century collectors] understood it to be a repository of the collective imagination of their society."[40] Museums have been conceived as a "common stock" of collections organized around a theoretical principle to educate the public.[41] Carla Yanni thus writes, "Since biologists tend to study organisms in the laboratory or in the field, they do not need to study museums as nineteenth-century paleontologists and taxonomists did. . . . Rather than admit their obsolescence, natural history museums have shifted their focus to educating the public about conservation."[42] What has changed, then, with the shift from prevention to preparedness in museums is that the value of collections comes less from their encompassing a hierarchical order of progress than from their exposure to a common disaster.

Avian reservoirs are thus inscribed in a more general historical trend, which Nelia Dias and Fernando Vidal have described as an "endangerment sensibility,"[43] characterized by "the expansion of museological forms of management."[44] In this new vision of the world, the value of living beings doesn't come from their accumulation for profit, which transforms them into standardized commodities, but from the imagination of future threats, which instantiates a list of priorities and scenarios for interaction. If the environmental value is not intrinsic but depends upon practices of collecting, storing, and classifying, as Dias and Vidal argue, then avian reservoirs create values through the perception of birds and the pathogens they carry.

While debates about preparedness often turned around its capacity to predict an event that never happens, I focus in this book on the infrastructures of collecting that made these predictions possible. Working in a museum, where public statements count less than the artifacts displayed, allows me to question the value of prophetic claims on bird flu not by focusing on the future events that they predict but by looking at the living beings they transform in the present. Anthropologists can work with microbiologists and ornithologists on the same footing when dealing with avian influenza because they share the same genealogy of Western museums—that is, a form of accumulation that produces knowledge by connecting a center with its peripheries. By tracing how these professions have built knowledges to anticipate the future, we can understand how they can curate the present—that is, display their objects in a critical way which is both meaningful and careful.[45] I now want to describe how the shift from prevention to preparedness affects the museums in which anthropologists think and work. In the previous two chapters I have asked how anthropologists have replied to demands of expertise about zoonotic diseases by framing the social as an ontological domain and as an ethnographic space; I now want to ask what kinds of practices of collecting and curating they have developed. If anthropology has thrived in Western countries in the last two centuries as a scientific practice connecting universities and museums, I will specifically consider how it has developed in France, where I have been trained to work as an anthropologist.

When the musée du quai Branly was created in 1996 by President Jacques Chirac, it merged two different types of collections: those of the former Musée de l'Homme (Museum of Man) and those of the Musée des Arts d'Afrique, d'Amérique et d'Océanie (Museum of Arts from Africa, America and Oceania). The genealogy of these two collections must be drawn separately to understand the problems raised by their fusion.

The Musée de l'Homme was created in 1938 by Paul Rivet and Georges-Henri Rivière with the former collections of the Musée d'Ethnographie du Trocadéro. It contained artifacts coming from gifts to the royal cabinets (mostly from America) and bounties of imperial expeditions (mostly to the Pacific), as well as skulls for republican phrenology (mostly from Africa). Specializing in South American cultures and languages, cofounder of the Institute of Ethnology of the University of Paris with Lucien Lévy-Bruhl and Marcel Mauss in 1925, Rivet promoted ethnographic missions through which objects would be collected, documented, and stored, such as the famous mission Dakar-Djibouti organized by Marcel Griaule between 1931 and 1933 or Claude Lévi-Strauss's expeditions in Amazonia in 1935 and 1938. Following the tradition of the French Enlightenment, the Musée de l'Homme presented itself as a repository where all cultures of the world could be studied through the ethnographic artifacts conserved in Paris. It was both a laboratory for the information contained in the artifacts and a space of exhibition for the aesthetic values of these objects.

Rivet allied with Rivière to display his collections following the new taste for "primitive art" developed by Surrealist artists. Both of them were key actors of the International Council of Museums (ICOM), a section of UNESCO that played after 1945 a similar role for anthropology to that of WHO for virology and IUCN for ornithology. Rivière was appointed director of ICOM between 1948 and 1965, and between 1937 and 1967 he became the head of the museum of European folklore named Musée des Arts et Traditions Populaires.[46] Rivet was one of the strongest supporters of the role of museums in public education and scientific research at the ICOM, although his socialist commitment and his faith in human progress were tarnished at the end of his life by his refusal of the independence of Algeria.[47] He thus declared at a meeting of ICOM in Paris on November 8, 1947, "UNESCO contributes by the available means—press, lectures, radio broadcasts, etc.—to the creation in the world of public establishments consecrated to develop science, to spread scientific culture, to diffuse the importance of research and scientific discoveries and their results on human progress."

The Musée des Arts d'Afrique, d'Amérique et d'Océanie was created in 1969 by André Malraux, who thus concluded his ten-year support of General de Gaulle after the communist anticolonial engagement of his youth. Under the presidency of De Gaulle in 1959, Malraux had created a minister of culture separated from the minister of education, arguing that education was addressed to the minds while culture was addressed to the hearts. He promoted the idea of an imaginary museum whereby all cultures of the world would communicate affectively, in rupture with the structuralist conception of cultures developed at the same time by Claude Lévi-Strauss at the Collège de France.[48] The ethnographic collections displayed at the Porte Dorée, on the site of the former Museum of Colonies, were supposed to trigger universal feelings in the visitor, such as fear of the sacred or the desire for human origins.

Placed under the double sponsorship of the minister of culture and the minister of research, the musée du quai Branly results from the tension between these two perspectives on ethnographic artifacts. With its 300,000 objects and 500,000 documents, it is one of the biggest museums of "non-Western art" or "world cultures," but it escapes the framework of other museums of anthropology. Most anthropologists, following the prominent voice of Louis Dumont, criticized the creation of the quai Branly because Jacques Chirac claimed that the idea of a "museum of primitive art" had been inspired in him by a collector and merchant of African art, Jacques Kerchache.[49] Claude Lévi-Strauss gave his support to this project because, he argued, the Musée de l'Homme had not secured properly the conservation of ethnographic collections, and many artifacts had been stolen and sold on the art market. He declared in 1992, "I have always said and thought that museums are made first for objects and then only for visitors, even if I go against trendy ideas. The first function of museums is conservation."[50]

To fulfill this function in a contemporary way, the musée du quai Branly invested in two major technological innovations: securitization and digitalization. Architect Jean Nouvel conceived an open storage at the entrance of the museum for musical instruments and a closed storage underneath the museum for all the other artifacts. The closed storage is accessible to only two of the four hundred persons working at the museum; those two use digital recognition to open its doors. All the artifacts acquired or lent for temporary exhibitions enter the reserves through a room where oxygen is removed by a technique called *anoxia* in such a way that the insects they contain die in a few days. The reserves are surrounded by a clay shell, which is supposed to protect the artifacts from a flood of the river Seine. The digitalization of collections reverses this politics of restricted access. All artifacts of the collection are available on the Internet through a software called TMS (The Museum System), with documents on their origins and composition. Visitors who want to have access to the objects themselves have to request permission from the curators a few weeks in advance and receive an appointment at the "muséothèque." Virtual images of objects thus compensate for the difficulty to access the real objects themselves.[51]

The exhibitions displayed by the museum reflect this ambivalent politics of access. In the permanent exhibition, called "plateau des collections," the masterworks of the collection are displayed in the dark with little information on labels or screens to let the visitor experience the mystery of these intriguing objects. Temporary exhibitions then shed light on some aspects of these collections by associating them with historical, aesthetic, or philosophical arguments. If visitors have indirect access to the objects under a glass box in the exhibitions, these exhibitions play on different forms of lighting—in the material and intellectual sense—to trigger reflections on the diversity of cultures. While the permanent exhibition is organized as a "tour du monde," no general narrative is proposed that would bring definitive coherence to the diversity of artifacts. The musée du quai Branly presents itself as a polyphonous expression of globalization.

This staunch refusal of the evolutionary narrative that was shared by anthropologists of the Musée de l'Homme and this assumed aesthetic of darkness inherited from the Musée des Arts d'Afrique, d'Amérique et d'Océanie—leaving antique Asian collections belonging to the Musée Guimet on the other side of the river Seine—mark a shift from prevention to preparedness. Ethnographic artifacts are not displayed in a narrative of progress where they would become meaningful as rare documents of the human species. They are considered as vulnerable materialities whose value has increased on the art market and that must be conserved for their scientific, aesthetic, and financial value. The musée du quai Branly faces not the threat of disappearance of the societies these objects come from, as the former Musée de l'Homme based on a scenario of evolution from primitive to civilized societies, but the threat of destruction of the objects themselves and of the different values they contain, based on the scenario of extinction of the ecologies in which they are conserved.[52]

Ethnographic artifacts are mostly composed with organic materials (wood, skin, skulls, feathers, saliva . . .) that can easily be degraded by their exposure to bacteria or to insects. Preventive conservation, which has been designed to calculate the time of degradation for the materials of classical European art such as wood or stone, thus needs to adapt to materials that were not conceived for storage in a museum. If ethnographic artifacts have been preserved beyond their expectancy of life by technologies, discussions on the duration of preservation involve political choices on the value of artifacts considered as national heritage and benefiting from these costly technologies. This is one of the main arguments in the debate on restitution of ethnographic artifacts to the societies they come from: they should be presented to a large number of visitors to justify the investment in their conservation. As artifacts are lent to other museums, an assessment of their state of preservation is made prior to the lending in order to calculate the insurance cost of the lending; but insurance cannot cover catastrophic events such as their massive destruction by a natural disaster. As contested and vulnerable objects, ethnographic artifacts need to be displayed and secured at the same time to continue their social life

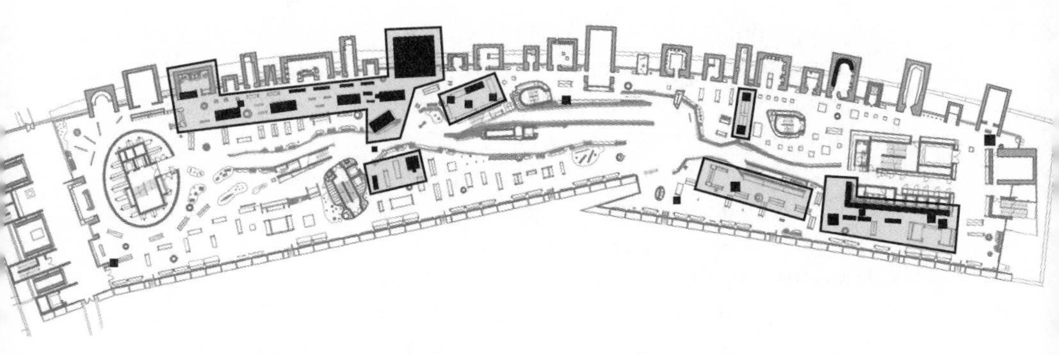

FIGURE 3.1. Map of the risks of infestation in the main exhibition space of the musée du quai Branly. Copyright musée du quai Branly.

as things.[53] While James Clifford defined museums as "contact zones" to characterize their openness to fluxes of persons and goods in a postmodern world,[54] they tend to be redefined as "spaces of contagion" by their circulation in the world of biosecurity.

Techniques of preparedness such as sentinels, simulations, and stockpiling have been applied at the same time in the collections of the musée du quai Branly in Paris and in the bird reserves of Hong Kong at the end of the twentieth century. The open storage of musical instruments at the entrance of the museum has been defined as a "hotspot"[55] of infestation after a map of the presence of insects was built using a technology to record the ultrasounds they make when they crunch the wood; the recording is called "Requiem for Xylophages." This map showed that the musical instruments near the cloakroom were more exposed to fleas carried by visitors. Musical instruments, with their low value on the market of tribal art, can thus be defined as sentinels for the threats that affect all the artifacts of the museum.

If infestation is a constant and minor threat as it slowly erodes wooden objects, flood is an exceptional threat that is much more dreaded, because it could damage all objects composed of organic material as well as documents. Since a major flood of the river Seine is expected every century, exercises of evacuation of the collections are organized every year, once for documents and once for artifacts. The staff of the museum is keen to participate in these exercises because it is one of the only opportunities for them to have direct access to the artifacts, which they move to the exhibition spaces as potential victims of disaster in need of care and support. Aside from these real-ground exercises of natural disasters, the musée du quai Branly also creates scenarios for digital applications enabling visitors to interact with objects: through enhanced reality, objects tell their stories from their creation to their arrival in collections, as if to exorcize the potentiality of their restitution.

In its program of contingency planning, the musée du quai Branly has classified its artifacts by order of priority for evacuation in case of a flood. Surprisingly, this order is not determined by the vulnerability of the materials to water exposure, but by the value of the artifacts on the art market. Thus, an expensive wooden statue from Africa would be ranked first for evacuation, although it could certainly resist flood for a few hours, while a series of botanical herbs from Vietnam or sugar skulls from Mexico would be ranked third because they have a lower value on the art market. In that sense, we can describe the reserves as a form of stockpiling and compare objects to vaccines depending on the strength of the immune response they trigger in those who observe them.

As an anthropologist working in the research department of the museum, I was in charge not of conserving or documenting ethnographic objects, but of reflecting on their conditions of production and display. How, I asked, have ethnographic artifacts shifted from objects of prevention to objects of preparedness? How did the ecologies of conservation mutate from a paradigm of evolution to a paradigm of extinction? And how has this shift in the pastoral management of their collections allowed curators to experience the meanings and uses of artifacts in hunting societies where most of them were made?

I have thus moved from an anthropology of global health to an anthropology of global art—two sides of the contemporary ecologies of conservation. If techniques such as sentinels, simulations, and stockpiling shift from the calculation of risks to the imagination of a disaster, my task as an anthropologist is to describe this imaginary and question how it can transform our ecologies of conservation. In the same way as I don't document ethnographic collections in a museum (curators do it better than I do with their training in art history), I don't investigate the risks of being infected in a farm, a lab, or a wildlife reserve (epidemiologists do it much better than I do with their training in statistics). Using classical methods of anthropology, combining ethnographic description and theoretical reflection, I trace how images emerge, act, circulate, and become exposed when humans interact with birds and viruses at the occasion of public health crises.

In the previous chapters, I have presented the history of avian reservoirs from a Western perspective: how pathogens passing from birds to humans have changed the way animal diseases have been problematized and occasioned alliances between microbiologists and birdwatchers. I now turn to more ethnographic analyses of techniques of preparedness from an Asian perspective: how have societies such as Hong Kong, Taiwan, and Singapore reflected about their position as sentinels of avian influenza? How is the idea of preparedness at the avian level inscribed in Chinese modes of perception of future events and of relations to birds? How do techniques of preparedness, with their different genealogies in Western science and global health, work in specific Asian territories?

techniques of preparedness

PART II

CHAPTER FOUR

sentinels and early warning signals

In this chapter, I describe how birds are used as sentinels—that is, how they are made to signal threats affecting humans. The idea that birds carry signs of the future that humans should learn to read is ancient, going back to Roman divination.[1] But what is new in the recent use of sentinel birds is the idea that signs sent by animals in the wild are analogous to signs observed between cells in a laboratory. The notion of cell signaling, derived from immunology, has led to identifying a complex system of pathways through which the organism builds a frontier between self and other. But this complexity also opens the possibility of being lured by wrong signals, which we have encountered in chapter 2. How is the practice of hunting and collecting viruses from birds reconfigured at the time of immunology, and how does it shape the trust in signals sent by sentinels? I will start with an ethnography of sentinel chickens in Hong Kong and then move to different levels where the question of lure and trust can be raised.

SENTINEL CHICKENS

In the summer of 2009, I was working on a poultry farm in Ha Tsuen, close to the city of Yuen Long in the New Territories of Hong Kong. The farm had been infected with H5N1 six months before and had received visits from microbiologists, journalists, and an anthropologist—I was the only one who stayed. The owner of the farm, Wang Yichuan, was also the head of the Hong Kong Poultry Farmers Association. This trade union was founded in 1949 with 145 farms breeding around 1,000 chickens. After the government issued a Voluntary Surrender Scheme in 1997 to encourage farmers to retire, there were only twenty-nine licensed farms in Hong Kong raising around 50,000 chickens each.

Wang Yichuan portrayed himself as a modern poultry farmer. He communicated regularly with the media, who described his enterprise as a "model farm" (*mofan nongchang*). A former truck driver, he had bought this farm in 1994 after he read in the newspaper about a Chinese man from Singapore who became rich by raising chickens. His wife, originating from a poultry-breeding family, advised him not to engage in this business, which she found tiring and risky. Wang's first difficulties came from the treatment of the waste, which triggered complaints from the neighborhood. Then, after 1997, they came from bird flu. It was a common occurrence to find ten dead chickens per day, he said, but when he found 200 on December 6, 2009, he realized that there was something wrong. Even more alarming, half of these dead poultry were "sentinel chickens" (*shaobingji*).

In Chinese, *shaobingji* literally means "whistling soldier chicken." A sentinel is a soldier who sends signals from the advanced posts of the battlefront.[2] These hundred "sentinel chickens" that had died in Wang Yichuan's farm had purposefully not been vaccinated for H5N1. Their deaths thus meant that the farm had been infected by an avian influenza virus that was transmissible to humans. Here, chickens appear as allies to humans in the war against flu viruses, because they are the first to die on the frontline. The word *sentinel* captures well the mix of agriculture, public health, and military concerns that constitute biosecurity measures and that can be also perceived in expressions such as *Voluntary Surrender Scheme* and *model farm*.

In 2003, following two outbreaks in the district of Yuen Long, the Hong Kong government had imposed a compulsory vaccination of all poultry raised on the territory. All farmers were required to vaccinate their chickens with a vaccine containing an oil emulsion adjuvant and a killed H5N2 an-

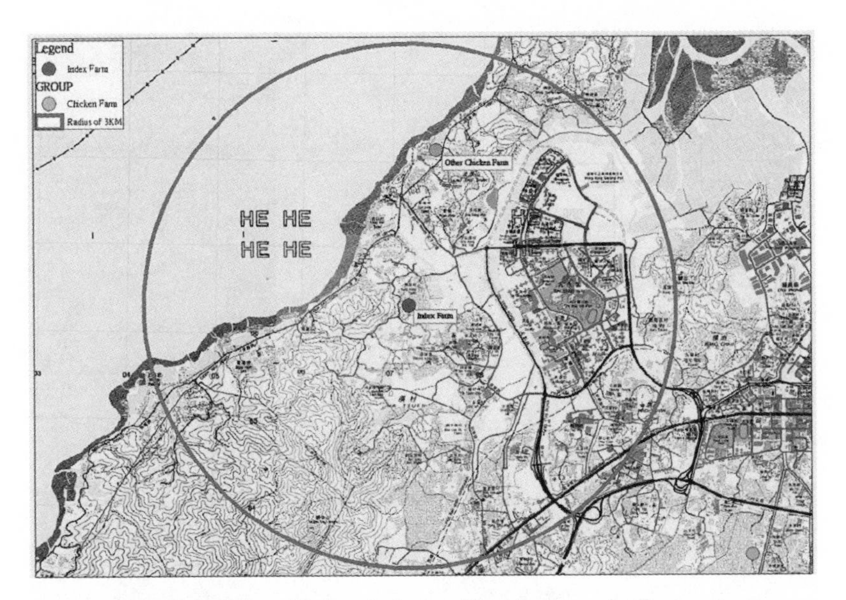

FIGURE 4.1. Map of Wang Yichuan's farm in Yuen Long, and of its location in the Deep Bay Area. The circle draws the security perimeter around the outbreak. Courtesy of the Agriculture, Fisheries and Conservation Department (AFCD) of Hong Kong Special Administration Region (HKSAR).

tigen, produced by Intervet in the Netherlands. Chickens were vaccinated at nine to eleven days old, with a booster four weeks later, and another booster at 150 days of age for broilers selected as future breeders. Unvaccinated chickens were placed at the ends of the rows of cages, with a rate of sixty sentinels for a flock of 3,500 broilers or 500 breeders. Staff of the Agriculture Food and Conservation Department of the government visited farms every week to check that vaccinated chickens were immune and that sentinel chickens were pathogen free.

The use of sentinel chickens in farms is not restricted to avian influenza. They are also applied to Newcastle disease, which affects only poultry and has very damaging effects.[3] But they can also warn about a disease that causes more damage in humans than in birds, such as Ross River virus and Murray Valley Encephalitis virus in Australia. Chickens are put in cages all over a territory to be bitten by mosquitoes and then sampled to check if they seroconvert.[4] Wild birds can also be considered sentinels when they die massively from an emerging disease, such as the West Nile virus in the United States.[5] If symptoms of diseases are very different in birds and humans (for instance, chickens infected with highly pathogenic avian influ-

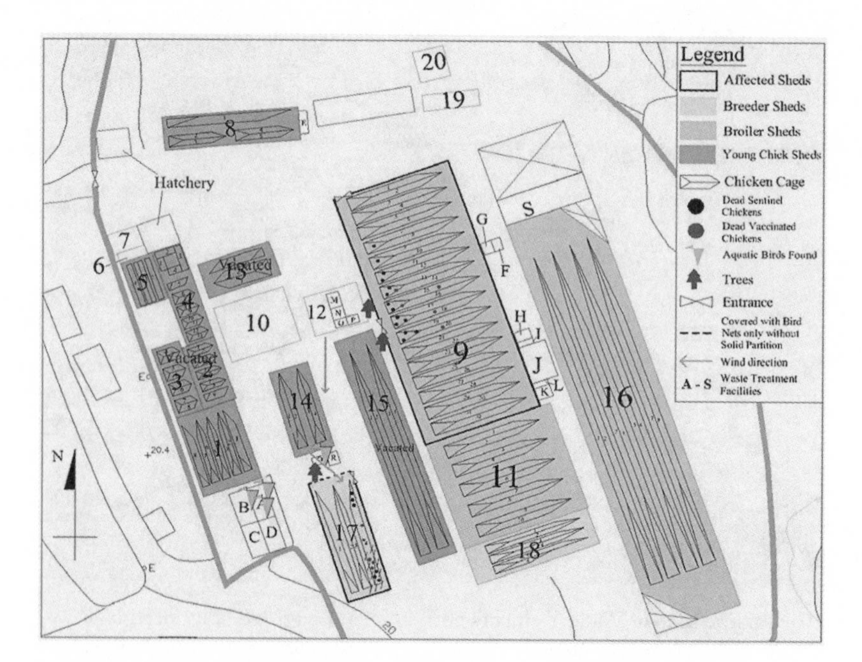

FIGURE 4.2. Map of Wang Yichuan's farm in Yuen Long. Courtesy of the Agriculture, Fisheries and Conservation Department (AFCD) of Hong Kong Special Administration Region (HKSAR).

enza are struck in their digestive tract, which becomes a bloody pulp, while humans are impacted in the upper respiratory track, causing pneumonia), they can still clearly signal from one species to another. If birds seem less worthy of human empathy than cows (although there is a form of communication with birds through sharing their songs, as we will see later), they are easier to equip as sentinel devices, since they are smaller, more numerous, and more mobile. Canaries were trapped in cages and placed in coal mines during the nineteenth century to warn of the release of dangerous gases because of their capacity to absorb large quantities of oxygen and thus detect toxicity in the atmosphere before it affects humans.

After the death of sentinel chickens was reported to the government, a team of experts was sent to Wang Yichuan's farm to investigate the cause of the outbreak. This team was led by Yuen Kwok-Yung, head of the microbiology department in Hong Kong University and of the Scientific Committee on Emerging and Zoonotic Diseases at the Centre for Health Protection of the Hong Kong government Department of Health. After working as a surgeon for the Hong Kong police department, Yuen Kwok-

Yung had been the first to report the cases of highly pathogenic H5N1 in 1997.[6] He was also a strong supporter of biosecurity measures in farms and markets in order to maintain the tradition of eating freshly killed, local poultry while protecting the population from avian influenza.[7] In a picture on the cover of the *South China Morning Post*, Hong Kong's main daily newspaper in English, Yuen Kwok-Yung was portrayed bending over Wang Yichuan—the virus hunter above the pastoral caretaker.

Several rumors were circulating in the media. Some said that the virus had arrived through smuggled eggs from mainland China, others that it came through wild birds, while some blamed a failure of the Dutch vaccine. The team of experts had to put these rumors to the test of virology. As the first markers of the presence of the virus on the farm, sentinel chickens were the starting point for the investigation. The mortality rate in the breeders' shed was 83 percent in sentinel chickens and 7 percent in the rest of the flock; in the adjacent shed, it was 43 percent for sentinel chickens and 0.2 percent in the rest of the flock. Breeders are more susceptible than broilers to infection because they have more contact with workers through regular insemination, while broilers never exit their cages. Around 2,500 swabs from neighboring farms were also collected on sentinels and vaccinated chickens, and those tested negative. Wild bird species previously found with H5N1 in Hong Kong, such as egrets and sparrows, were seen on the farm, but all the samples from them tested negative as well.

The samples collected in the two infected sheds showed that the H5N1 virus belonged to Clade 2.3.4, which was circulating in poultry and wild birds in south China and Vietnam. Consequently, the expert team concluded that they could not identify the exact source of the virus; it could have been people entering the farm, wild birds, or even the wind. Smuggled eggs were excluded, as the farm produced its own one-day chicks. The team could say that the virus had been amplified by two factors. One was the lack of respect ("inconsistent observance") of biosecurity measures: workers did not wear masks or gloves and often forgot to clean their boots in the ponds at the entrance of the sheds; the manager mixed the disinfectant with bleach in the ponds; nets had holes that sparrows could go through. The other factor was the concentration of sentinel chickens at the end of the rows of cages, which produced a heavy viral load that challenged the immunity of vaccinated chickens; this meant, paradoxically, that an excess of biosecurity could work as an amplifier of the virus. It was thus recommended the farm should increase biosecurity measures and scatter sentinel chickens among the vaccinated chickens.[8]

Wang Yichuan complied with the biosecurity measures the government required. After the outbreak, his six workers were sent to the hospital to check their health, his two daughters were sent to their grandparents' house, and he stayed with his wife in his farm under quarantine. He had to kill and bury the 70,000 chickens living on the farm as well as 25,000 fertilized eggs, clean all the equipment, and change the nets. "There is not one feather left,"[9] he proudly told me. When I worked at his farm in August 2009, he raised 30,000 chickens (his license allowed him up to 100,000) and only employed four workers; but he had bought another farm in mainland China, where he employed four other workers, and he was regularly crossing the border for his business. "Bird flu is a risk I am ready to take," he told me, "because I like the poultry business. You can lose a lot, but you can also earn a lot!" He talked to me about the spread of Chinese merchants around the world, from the army of Genghis Khan to contemporary businessmen in China. He insisted on speaking *guoyu*, the national language, rather than *guangdonghua*, the local dialect. When I asked him about the rumors on the cause of the outbreak, he said he understood that journalists needed to "add salt and vinegar," but that he preferred to talk in a straightforward and transparent way (*touming*). While other farmers spread the rumor that his farm had been infected by smuggled eggs, he firmly believed that sparrows had introduced H5N1.

His colleagues at the Hong Kong Poultry Farmers Association gently joked with him that he behaved "like a good pupil." The solidarity between poultry farmers faced with the outbreak had not been high: Wang Yichuan told me he did not receive much help cleaning the farm. When most farmers discussed fears about bird flu, they mentioned conditions of life in Hong Kong, which they ironically compared to the stressful life of chickens. "Nobody in the world has reacted like the government of Hong Kong," one of them told me. "All living beings have diseases. Chickens die, but they don't necessarily have a virus. In Hong Kong, people are stressed because they live in concentrated buildings, like chickens in a cage."[10] Yuen Long, a former village that became a middle-class town through the construction of public housing estates in the last twenty years, had grown to half a million inhabitants and had high rates of immigration, unemployment, and suicide. It was known as the "town of sadness."[11]

The workers he employed were all coming from mainland China. I spent most of my time with a young man from Guangzhou, Li Qigui, who was in charge of feeding chickens and cleaning their feces. Every day, we took the cornmeal (also coming from mainland China) from a huge silo and brought it to the mangers on the side of the cages, under the strong airflow of huge ventilators. Cages were piled one on the other in a scale, which allowed the worker to bring the food in a descending way, but also allowed chickens to defecate in the voids below cages. In the afternoon, we had to clean the huge quantity of fecal matter brought by mechanical shovels on the side of the cages. Using manual shovels, we had to put it into bins that were emptied by trucks every two days. Sometimes, dead chickens were put in plastic bags and thrown in these bins. This work allowed me to see a poultry farm as a metabolic factory producing both living material and fecal waste in almost equal quantities. It was therefore tempting to imagine all the viruses contained there.

I also spent some time with Yan Yuren, an elderly woman who had left her whole family in Fujian five years earlier to work on the farm. She was taking care of the young chicks and the breeders, giving them drugs and vaccines through a long needle (*dazheng*) that intrigued Li Qigui; he asked me questions about it, because he thought I was an expert on bird flu. Yan Yuren showed me the sentinel chickens at the end of each row and told me that they were sent to the market if they were in good health. She also said that she could not scatter them as required by the government, because she would not know which was vaccinated and which was not. She asked me to feed the chicks with a bag of enriched seeds and to first remove all the small insects (*zhong*) contained in the feed by passing them into a filtrating machine. *Zhong* is a common name for bugs in ancient China, and it seemed that insects were a bigger concern for Yan Yuren than viruses.[12]

Yan Yuren also cooked food for us and for the other workers—two middle-aged men who were renovating the cleaned sheds. I was surprised to see that she prepared pork, fish, vegetables, and rice, but never chicken. When I asked if they ate chickens, she asked Ms. Wang to kill one; she regularly did it "for friends," she said. We had the chicken for dinner, after taking a shower and changing our clothes, as if eating chickens implied breaking with their daily routine.[13] To emphasize the rupture with ordinary meals, I had brought a bottle of red wine from France, which my coworkers considered with perplexity. They finally drank it in their bowls with the chicken as a kind of soup at the end of the dinner. I later realized that Chinese wine (*maotai*) is white and can be cooked with chicken, while red wine looked like blood to them. As in Lévi-Strauss's classical analysis, wine was a sign of the conditions that made this moment of reciprocity possible, the killing of chickens for food production and the sharing of their blood, which explains its separation from other consumable goods.[14] We finished

FIGURE 4.3. Yan Yuren giving vaccines in a chicken farm in Yuen Long. Photo by Frédéric Keck, August 2009.

the *maotai* after the meal while watching a series on TV about the Chinese People's Liberation Army.

At the end of every week, a couple of bird-catchers came at night to take around 1,000 chickens to the wholesale market. The chickens were removed from their cages while they were sleeping and then stacked in red cages in stressful conditions. Their feathers and feces had to be cleaned from the ground after the catcher passed by, and the truck was cleaned with water at the end of the operation. One night, I drove with the catchers to the wholesale market. On the way, they talked to me about American universities, where they wanted to send their sons. They seemed to have a good standard of living as necessary intermediaries in a profitable industry. When they arrived at the wholesale market, the red cages were placed on the ground to wait for the opening of the auctions at 6:00 A.M. Other trucks arrived from mainland China with live chickens in yellow cages. They had been inspected at the Sheung Tsui border and were once again checked and cleaned at the wholesale market. Every day, around 10,000 chickens arrived from mainland China, and the same number arrived from

FIGURE 4.4. Chickens are taken from the farm in Yuen Long to the market in Cheung Sha Wan. Photo by Frédéric Keck, August 2009.

Hong Kong farms. But the price of a chicken from Hong Kong was twice that of a chicken from mainland China (around 60 Hong Kong dollars versus 30). They were then sent to retail markets all over the territory, where they were sold alive to consumers.

When we left the village of Ha Tsuen in the truck, we passed containers waiting to depart on the banks of the Pearl River Delta. A motto, "We carry, we care," was written on these containers. It seemed to capture Hong Kong's vocation as the gateway for commodities coming from China: carrying goods to the market was linked with a care for their safety. The concern for public health during outbreaks of bird flu, with the oppositions it draws between a good, edible chicken and a bad, infected chicken and between a profitable commodity and a dangerous living being, masks the caring relationships in the production of livestock that makes consumption possible.[15] Sentinel chickens, used as markers of extraordinary outbreaks and yet consumed as ordinary chickens, were emblematic beings in that secured food chain: neither mere commodities nor pure living beings, they sent signs of coming threats and future goods.

SENTINEL POST

Like sentinel chickens at the end of a row of cages in a poultry farm, Hong Kong had been charged with being a sentinel post for pandemic avian influenza. Recall the quote from Hong Kong University microbiologists that I mentioned in the introduction: "The studies on the ecology of influenza led in Hong Kong in the 1970s, in which Hong Kong acted as a sentinel post for influenza, indicated that it was possible, for the first time, to do preparedness for flu on the avian level."[16] What does it mean for the whole territory of Hong Kong to be a sentinel post for bird flu at the global level?[17] How does it express the relation between Hong Kong, China, and the rest of the world? I want to draw an analogy between sentinel chickens, vaccinated chickens, and the manager of the farm on one side and Hong Kong citizens, their Chinese neighbors, and their government on the other to investigate modes of identification between humans and birds at this political level through the anticipation of an avian influenza pandemic.

The central market of Cheung Sha Wan was situated in the middle of Kowloon, one of the densest areas in the world. When I heard about the H5N1 outbreak in December 2008, I went to Cheung Sha Wan, where I was able to take pictures of the live chickens being gassed under the gaze of the journalists. These 10,000 chickens came from a farm within three kilometers of Wang Yichuan's farm and were killed as a precautionary measure.[18] Given that another farm close to Wang Yichuan's also had to kill 18,000 chickens and that Wang Yichuan had to kill 70,000 chickens on his farm, this means that 100,000 chickens were killed as a consequence of this outbreak. Killings in the central market were the public face of a massive slaughter that was taking place on the farms. During the previous H5N1 outbreaks, in 2001 and 2002, the numbers of chickens killed were 1.2 million and 900,000, respectively, which gives an indication of the decrease in poultry breeding in Hong Kong. But the first massive killing occurred in November 1997, after the first cases of H5N1 in humans and birds: between 1.3 and 1.5 million poultry were killed. A team of civil servants from the Agriculture Department was assigned to the regular practice of what the authorities called "culling." "Most of them had never seen live poultry before. They had to learn. Now some of them have become experts in poultry culling," the head of the Agriculture Department declared.[19] Culling, by contrast with killing, means the removal of sick parts to enhance the health of the whole—a practice borrowed from gardening.

The first massive slaughter of chickens as a precautionary measure against avian influenza occurred in 1983 in Pennsylvania: 17 million birds were killed because of a highly pathogenic H5N2 virus, which was not transmissible to humans but spread rapidly in an area of intense poultry breeding.[20] In 1995, Robert Webster, "the pope of influenza," was quoted in the *New Republic* as saying, "The chicken population in Pennsylvania [in 1983] is like the world as it is in this moment. What would we have done if this virus had occurred in humans? There are millions of us 'chickens' just waiting to be infected." The author of the article, Malcolm Gladwell, stresses the problems raised by this analogy: "Human beings do not live their lives crammed next to each other in metal cages. They do not wallow in their own feces. They have the brains and the ability to take precautions against disease and contagion. People are not chickens. So why all of a sudden are we convinced that we are?"[21] However, this analogy between chickens and humans was often made by the people I interviewed in Hong Kong about the fear of avian influenza, and it may have been used by the Hong Kong government when it faced the challenges of the mid-1990s.

Margaret Chan was the head of the Hong Kong Government Health Department in 1997. She resigned from this position in 2003 and became head of the World Health Organization in 2005. She is remembered by Hong Kong citizens for going to poultry markets under the gaze of televisions and saying, "I eat chickens, you can eat them too!" One of her advisors I met in Geneva recalled a conversation with her in a crisis meeting in 1997. She said, "Kill all the chicken!" He replied, "But what if there is still some virus?" "—Then kill all the ducks!" "—But what if it doesn't work either?" "—Then you will have to kill me!"[22]

This quote strikingly recasts the role of the emperor in times of crises, which often coincide with regime changes or revolutions (*geming*). In this analogical conception codified by the Confucian tradition, the sovereign or one of his or her representatives has to pass the test of sacrifice, gathering the thousand beings (*wu wei*), whose proliferation constitutes the universe ("everything under heaven," *tianxia*), and killing ritual animals (pigs, chickens, and cows) at the center of the political space.[23] For Chinese authorities, bird flu (*qinliugan*) comes from the intense trafficking of humans and animals (*renliu wuliu*), where *ren* means both *human* and *virtue*, *wu* means both *animals* and *beings in general*, and *liu* means both *flows* and *flu*. Every year, when Chinese migrant workers (*liuming*, floating population) return to their hometown from the big cities in which they work, authorities fear that they spread contagious diseases (*ganbing*).[24]

For Hong Kong citizens, the massive killing of birds in 1997 produced ambivalent feelings of relief and fear. As the Chinese People's Liberation Army had just entered Hong Kong to mark Beijing's sovereignty over the former British colony, the slaughter of potentially infected chickens could be seen either as a gesture of pastoral protection for humans or as a sign of a future political threat sent by animals.[25] While Hong Kong citizens feared being crushed by the Chinese army, chickens were massacred by their own agriculture authorities. A Chinese saying goes, "Kill chickens to warn monkeys" (*sha ji jing hou*), which indicates that the massive killing of chickens was also a sign of China's restored sovereignty over Hong Kong. The killing of more than one million chickens may also have recalled Mao Zedong's 1958 mobilization of the Chinese population against sparrows, which were considered pests.[26]

Many Hong Kong citizens had fled the Chinese People's Republic in the previous fifty years to work under hard conditions in the British colony, and backyard poultry were for them both companions and sources of proteins. In 2006, raising backyard poultry in Hong Kong was forbidden by the government to prevent the spread of bird flu. But Hong Kong citizens of all generations remained attached to the consumption of freshly killed poultry raised on their own land, despite the government's encouragement to buy chilled poultry imported from mainland China. They kept going to live poultry markets (also called "retail markets" or "wet markets") where chickens were killed in front of customers who could thus check their health.[27] The exceptional massive killing of Hong Kong chickens, despite its proclaimed goal of caring for the citizens, was thus perceived as a destruction of long-time attachments between people and backyard poultry and as a sign of Hong Kong's increased dependency on mainland China, while the daily ordinary killing in live poultry markets paradoxically maintained their proximity with poultry.

The rationality of this sacrificial gesture, however, was shared by most of the actors who contested its economic and ethical consequences. The Hong Kong Buddhist Association prayed for the souls of dead chickens all along the borders of the territory. Rather than killing the chickens at the center of the territory, the Buddhists let their souls escape in a way that would reduce bad *karma* for their next life. A practice called *fangsheng*, literally "release life" or "let live," when Buddhist practitioners bought birds in markets and released them in surrounding natural parks, had to be forbidden by the government, because many released birds were found dead and some were infected with bird flu. The Taoist *jiao* festivals organized in villages of the New Territories also had to stop killing roosters at the beginning of the three-day vegetarian celebration of the divinities of the environment. The killing of the rooster, whose blood was spread at the borders of the village, was traditionally meant to evacuate bad energies from the site of the festival.[28]

The Buddhist and Taoist practitioners I met explained the emergence of H5N1 as bad energies or bad karma produced by an increasing consumption of chicken meat and said that chickens were taking revenge against humans by sending pathogens. While the Hong Kong government was supporting the livestock industry and meat distribution system that these religious groups contested, Buddhists and Taoists shared the same view of Hong Kong as a bounded territory that had to be protected from outside threats. If they criticized what they perceived as a Confucianist sacrifice to restore the power of the sovereign in a period of regime change, they used the same analogical ontology, in Descola's terms, or to borrow Foucault's description of sovereign power, the same techniques of "make die and let live."

Hong Kong microbiologists, by contrast, developed a different perspective on avian influenza. Although most of them were not born in Hong Kong but in Australia, Sri Lanka, and mainland China, they reconfigured Hong Kong's new identity as a sentinel post under Chinese sovereignty. The Department of Microbiology of Hong Kong University (HKU) was created in 1972 by Kennedy Shortridge, a colleague of Robert Webster at the University of Melbourne and in the WHO expert committee on the ecology of influenza at Geneva. Since the 1968 pandemic flu virus, which killed around one million people worldwide, was first identified in Hong Kong and therefore called the "Hong Kong flu" virus, Shortridge and his colleagues expected that the next pandemic would also emerge from south China. The People's Republic of China, not being a member of WHO, did not share information about the flu strains circulating on its territory and did not consider influenza to be a major public health issue. Shortridge built networks of personal relations (*guanxi*) with veterinarians in Guangdong and collected samples of flu viruses among ducks and pigs in the area. He then proposed that the ecology of south China—where ducks are used as pesticides in rice paddies, in close proximity with humans and pigs—made it an "influenza epicenter" for the rest of the world. "The densely populated intensively farmed area of Southern China adjacent to Hong Kong," he wrote with the renowned British influenza expert Charles Stuart-Harris, "is an ideal place for events such as interchange of viruses between host species."[29] To support this hypothesis, he remarked that the Chinese character for house (*jia*, 家) contained a pig under a roof, as if one could see the mutations of viruses coming from animals when looking at the various traits of this traditional character.

When the first cases of avian influenza appeared in February 1997, Shortridge was sent to markets to raise the alarm. There were 1,000 live poultry markets in Hong Kong at that time, and in some of them 36 percent of chickens tested positive for H5N1. He recalled, "One moment birds happily picked their grains, the next they fell sideways in slow motion, gasping for breath with blood slowly oozing from their guts. I had never seen anything like it. I thought, 'My God. What if this virus were to get out of this market and spread elsewhere?'"[30] He later said that the dead bodies of chickens in the Hong Kong wet markets reminded him of his mother's description of the human victims of pandemic influenza in his native Queensland in 1918.[31] His strategy was consequently to find flu viruses in birds before they were attenuated by pigs so as to make a human vaccine before the virus jumped from pigs to humans. But in 1997, it was impossible to make a vaccine for a virus passing directly from birds to humans and lethal to poultry, since vaccines are cultivated on chicken embryos.[32] He therefore recommended the government have all the live poultry in the Hong Kong territory killed in order to eradicate the animal reservoir of H5N1.

"We didn't cull, we conducted a slaughter!" Shortridge told me in an interview.[33] When I asked him how the massive killing was acceptable to the Hong Kong citizens, he told me that five years earlier, he had recommended the closure of the horse races in Hong Kong because there was an outbreak of equine influenza, which is lethal in horses but not transmissible to humans.[34] In a city where horseracing provides the only opportunity to gamble and where the Jockey Club is the richest association, this closure was more costly to many Hong Kong citizens than the killing of their backyard poultry. Shortridge later justified the repeated killing of poultry infected with bird flu as a preemptive measure.

> Poultry were killed market-by-market as signs became evident, leading to the pre-emptive slaughter of all poultry to prevent human infection. Early detection and reaction was the order again in 2002 and 2003. Thus, there now lay the prospect for influenza-pandemic preparedness not only at the human level but, better still, at the baseline avian level with the ideal that if a virus could be stamped out before it infected humans, an influenza incident or pandemic will not result. In 1997, the world was probably one or two mutational events away from a pandemic, while in 2002, with earlier detection, it was probably three or four events away.[35]

We see here that Shortridge spoke in the language of probabilities, assessing the number of mutational events required for the virus to become pandemic. But his preparedness technique relied most heavily on imagination, as when he recalls his family memories of the 1918 flu pandemic or Chinese characters that endorsed his view of the "influenza epicenter." While Margaret Chan talked of precaution, in a pastoral rationality, Shortridge talked of preemption, in a military rationality.[36] Both of them mixed classical prevention and new techniques of preparedness that were emerging.

Shortridge's 1982 worst-case scenario—a pandemic starting with animals in south China—was later confirmed by SARS (Severe Acute Respiratory Syndrome). In November 2002, when the first cases of a mysterious pneumonia were detected in Guangzhou, thirty wild birds were found with H5N1 in Penfold Park, close to a horse track in the New Territories, followed by infected poultry in farms and markets.[37] While the first cases of SARS appeared in Hong Kong hospitals in March 2003, spreading airborne through emergency rooms, the HKU microbiologists lost two weeks because they were testing for an avian influenza virus and could not identify the pathogen causing this new disease. The SARS crisis was considered a public health failure in Hong Kong, since the disease rapidly spread through hospitals and middle-class housing estates. This explains why Margaret Chan stepped down from the Department of Health and why the department was quickly reorganized with a Centre of Health Protection dedicated to crisis management, separated from Hospital Authority.[38] However, SARS soon became a success story for the early detection of zoonoses.

This success was rightfully attributed to the activities of two HKU microbiologists, who were then presented as heroes in the public sphere for revealing the paths of SARS between humans and animals.[39] On March 18, 2003, Malik Peiris identified the virus that had caused the new disease by cultivating human samples on dog and monkey cell lines; he found it was a coronavirus, a virus with a big capsid that is usually benign but had become lethal by a mutation in its Spike protein.[40] He observed that this virus could survive for two days outside a cell and transmit by droplets. Peiris was born in Sri Lanka and trained in microbiology at Oxford; before moving to Hong Kong in 1997 to study avian influenza, he had conducted innovative research in his native country, collaborating with veterinarians on the spread of Japanese encephalitis through the displacement of pig farms, which had first led him to think about the role of livestock in emerging infectious diseases.[41] The other hero was Guan Yi, who was born in rural

Jiangsu, studied virology with Robert Webster at Memphis after 1989, and returned to Hong Kong in 1997 to "track down" the bird flu virus. Both described themselves as virus hunters and liked to imagine themselves as viruses invading cells: while Guan Yi played on his rural Chinese background to portray himself as walking through the floating populations of Guangzhou to invade Hong Kong,[42] Peiris explained that viruses crossing species borders act as strangers in an environment unfit for them—a situation he had met himself when he moved from Sri Lanka to Oxford and Hong Kong.[43]

In February and March 2003, thanks to his connection with the chief of the Guangzhou Institute of Respiratory Diseases, Zhong Nanshan, Guan Yi crossed the border between Hong Kong and mainland China to collect samples, first from hospitals, and then from wet markets in Guangzhou and Shenzhen. He could thus show in May that the coronavirus identified by Peiris was present in masked palm civets—a rodent cooked in Chinese traditional medicine to cure respiratory diseases—and in bats, which were increasingly migrating into cities from the forests of south China.[44] The animal reservoir of SARS was thus discovered: the coronavirus was circulating in bats and spread to humans by the "mixing vessel" of civet cats. The politics of massive killing was consequently applied to Chinese wet markets, and the consumption of civets was forbidden.

The SARS crisis thus confirmed—although retrospectively—the importance of the early detection of emerging infectious diseases. By contrast, it reinforced among Hong Kong experts the idea that in mainland China, the early signals of a threat were blurred. During the first steps of the crisis in winter of 2002, news came to Hong Kong that people in Guangzhou were buying vinegar to cure a mysterious disease, but there was no information about the disease until it arrived in Hong Kong through a doctor who had treated patients in Guangzhou and then infected around ten people at the Metropole Hotel in Kowloon on February 21. When the disease spread to the rest of China and the globe—Vietnam, Taiwan, Singapore, Philippines, and Canada—the Beijing government, engaged in the political turmoil of the seventeenth Congress of the Communist Party that was to officialize the shift from Jiang Zemin to Hu Jintao as the head of the country, refused to communicate information to WHO. On April 9, 2003, *Time Asia*, based in Hong Kong, published a letter by a Beijing physician who had treated victims of the 1989 massacre and accused the Chinese government of hiding its SARS cases.[45]

The global pressure on China forced the new government, led by Hu

FIGURE 4.5. "SARS heroes" at the Hong Kong Museum of Medical History. *From left to right:* Yi Guan, Kwok-Yung Yuen, John Nicholls, Malik Peiris, Honglin Zhang, Leo Poon. Photo by Frédéric Keck, June 2009.

Jintao and Wen Jiabao, to conduct a strong public health campaign against SARS, imposing quarantine, prescribing masks, and building hospitals all over the country. When the epidemic ended at the end of June 2003, it had infected 5,000 people in China, 350 of whom died, and 1,800 people in Hong Kong, with 300 dead. Despite the uncertainties on the Chinese statistics, the death toll in Hong Kong may have been higher due to the high urban concentration and the early reactions of panic, while Chinese traditional medicine used in mainland China seemed to work on the long-term effects of the disease.[46] If the SARS crisis had been longer, the Hong Kong economy would have gone bankrupt, since all flights and exchanges were suspended for a few months. The very assets that had made Hong Kong a "small dragon"—its dense, hard-working, and qualified population, as well as its position as a hub of transportation in East Asia—were now seen as vulnerabilities.

The strategy of HKU microbiologists was to turn this vulnerability into a new asset. The SARS crisis did a lot to transform Hong Kong from a commercial gateway to a public health sentinel post. Before the Second World War, Hong Kong was the *entrepôt* through which Chinese commodities were checked before being sent to the rest of the world.[47] During the Cold

War, despite the American embargo and the rise of a "made in Hong Kong" industry due to the arrival of migrants from mainland China working as a cheap labor force, it maintained this function.[48] In the 1990s, with the opening and modernization of the Chinese economy, it became a center for finance and trade, where contracts were passed between Chinese and Western laws. After 1997, due to the Asian financial crisis and the will of Jiang Zemin to turn Shanghai into the new commercial and financial center of China, Hong Kong's identity was severely questioned.

Bird flu and SARS were seized by Hong Kong elites as opportunities to present their position between China and the rest of the world as a site of early detection of public health and environmental threats. While the good hospital infrastructure was an effect of the welfare state policy implemented by the colonial regime as a response to the 1967 riots,[49] SARS led to a reinforcement of crisis management and biological research. The commemoration in 2004 of the discovery of the plague bacillus in 1894 by Alexandre Yersin gave Hong Kong an opportunity to reassert its role in the global research on infectious diseases as a two-century tradition.[50] When H5N1 spread to Southeast Asia, Japan, Russia, and Europe in 2005, the measures that had been applied in Hong Kong since 1997 were recommended to other countries by Margaret Chan, then head of WHO. During the 2009 "melamine crisis," when 300,000 babies in China were intoxicated by milk powder containing a chemical additive, the Centre for Food Safety in the Food and Environmental Hygiene Department of the Hong Kong government checked all the milk products on its territory and banned those that came from the mainland.[51] As Shortridge, Peiris, and Guan asserted in a previous quote from "The Next Influenza Pandemic," Hong Kong had become an experimental laboratory to apply preparedness at all points of the food production, and potentially of the environment at large.

This role as a sentinel post was inseparable from Hong Kong's reflection on its identity as a democratic regime in a neo-liberal economy. While the government of Tung Chee-hwa, a rich Hong Kong businessman chosen by Beijing authorities in 1997, was blamed for its lack of intervention during the SARS crisis, 500,000 people demonstrated in Victoria Park on July 1, 2003, just after the end of the SARS crisis, against an amendment to the Basic Law that would increase security measures and restrict civil liberties. Demonstrations against Beijing's sovereignty are organized every year on June 4 to commemorate the 1989 Tiananmen massacre; they climaxed between September and December 2014 when around 100,000 people occupied the streets of Hong Kong to protest against the Communist Party's control over the candidates for the election of the chief executive—an event known as the "umbrella movement." While the "Hong Kong flu" in 1968 had been a test for relations between the British government and the Communist Party after the massive strikes it had organized,[52] and while the emergence of H5N1 in 1997 has been a test for the new Chinese government to protect its population by almost eradicating its poultry breeding, the SARS crisis was a test for the "Special Administrative Region" under the "one country, two systems" rule that guaranteed free trade and public liberties for fifty years.[53]

How does this definition of Hong Kong as a sentinel post transform the relations of its citizens to birds? We have to go beyond the compassionate identification with the poultry "sacrificed" by the new Chinese sovereign, defining Hong Kong citizens as democratic and liberal citizens by contrast with their Chinese neighbors, to look at their more ordinary relations with birds on their territory. If Hong Kong is a sentinel for signs of future threats, it should not play this role only at the political level of the government regulating a territory, but also at the level of ordinary relations between citizens and their environment. After describing how virus hunters identify themselves with viruses when they track them through chicken farms and markets, we can now see how birdwatchers identify with birds they follow in their migratory flyways.

SENTINELS FOR THE ENVIRONMENT

When birds infected with H5N1 were found in 2005 on the shores of Qinghai Lake, in the northwest of China, and on the Russian borders of Europe, public health authorities called on ornithologists to answer the following question: can birds with flu fly? The Wildlife Conservation Society, a powerful international association based in the United States, developed an ambitious program with the international organizations in charge of human and animal health (World Health Organization, World Animal Health Organization, Food and Agriculture Organization) called "One World One Health," arguing that the management of avian influenza and other emerging infectious diseases needed to involve a global investigation on the interface between humans, animals, and their environment.[54] The alliance between environmental groups, virologists, and public health officials was one of the major effects of the emergence of H5N1. It produced new forms of sentinels, not only at the level of a farm or a territory, but also at the level of migratory flyways—a concept that was forged by an American ornithologist to describe the routes taken by birds between feeding sites on their migration from their summer to their winter habitats.[55]

Hong Kong is situated on the East Australasian Flyway, extending between Japan, Korea, China, Indonesia, and Australia. It is estimated that around 500 species are resident or migratory in its small territory—as many species as those observed in the whole European continent. This diversity consequently attracts birdwatchers from all over the world. The Hong Kong Birdwatching Society (HKBWS) was founded in 1957 by British officers who wanted to establish a list of the bird species on the territory for conservation purposes. Sir John Chapple, commander of British forces in Hong Kong and a friend of Governor Edward Youde, himself a passionate birdwatcher, was in charge of the surveillance of the Mai Po marshes. This territory at the end of the Pearl River Delta was a site of passage for Chinese refugees as well as for migratory birds. In the *Bulletin of the Hong Kong Birdwatching Society* in 2008, Sir Chapple thus explained how the military control of the territory became an ecological asset: "Military control meant control over access; military patrols seeking out illegal immigrants (which was the new classification of refugees) meant a constant oversight over the whole area, and this in turn helped to end illegal trapping inside the fence; better controlled road access allowed for useful environmental studies and also made easier the construction of board walks and hides."[56]

The British practice of birdwatching was thus linked to the military project of mapping and defending the colonial territory. The Mai Po marshes became a nature reserve when the government granted its management to the World Wildlife Fund in 1984.[57] The fishermen who cropped shrimps in the ponds (*gei wai*) were hired by the park to keep ponds at low tide and attract birds. With 10,000 migratory birds feeding in the marshes every year, it became a "Wetland of International Importance" under the Ramsar Convention in 1995. The first manager of the reserve, David Melville, had held the post of government ornithologist between 1974 and 1980. The WWF still trains Chinese conservationists from the mainland (particularly from neighboring Fujian) and teaches them how to protect wetland areas. It organizes the annual Hong Kong Bird Race, during which teams compete to see the highest number of birds in one day on the whole territory. While birdwatching was a leisure activity for the British colonial elites, it also followed the perceptual habits of a military army watching over the border with China, competing to build lists of numbers and training for a potential conflict.[58]

Hong Kong British birdwatchers also took part in the Migratory Animal Pathological Survey (MAPS), led by ornithologist Elliott McClure and Colonel Charles Barnes from 1963 to 1971 for the U.S. Army. This project involved capturing and banding birds throughout the Far East so as to assess the spread of Japanese encephalitis. It involved thirteen teams in nine countries and banded more than one million birds of 1,218 species.[59] While the headquarters of this project were based in Tokyo and Bangkok, Hong Kong was the reference center where tags found on birds all along the migratory flyway were to be mailed. This project was the first massive production of statistical data on East Asian birds, but also the first attempt to convert a military concern for biosecurity into an environmental care for biodiversity.[60]

For these two reasons, its situation at the end of the Pearl River Delta and its position in the middle of the East Asia Australian Flyway, Hong Kong thus became a sentinel post for environmental threats affecting birds and humans in common. But also, while avian influenza was reserved to experts working for the government, birdwatching turned Hong Kong into a democratic sentinel, involving an increasing number of citizens in the protection of the environment. After the 1997 handover of Hong Kong to China, the number of Chinese members of the HKBWS bypassed that of expatriates, rising to 1,500, which makes it the biggest environmental association in Hong Kong. The chair of the society was given to Lam Chiu Ying, then head of the Hong Kong Observatory, who became the first Chinese member of the society in 1976. When I met him, he described how he had made birdwatching a popular practice not just for the colonial military elite but for the new middle class that was discovering leisure and nature. "Before, they assumed that everyone had a car; I started running coach trips to Mai Po for members. Before, they thought birdwatchers should arrive at sunrise; I wanted the coach trip to start at 8:00 A.M. So we had more common people."[61] As an academic trained in the observation of stars in the sky, Lam Chiu Ying was a successful promoter of the wonders of birdwatching.

HKBWS's greatest success was the protection of the area of Long Valley, at the north of the New Territories, which attracted more members to the association. This agricultural wetland was threatened by a railway line planned by the Kowloon-Canton Railway (KCR). The HKBWS showed that Long Valley was home to more than 210 bird species and argued that many of them had disappeared from similar habitats on the other side of the border with mainland China. They launched a public campaign between 1999 and 2001 and secured the diversion of the railway into an underground tunnel that did not affect the habitat. Lam Chiu Ying recalled, "At that time, we had virtually no chance of winning since we were fighting the railway company which was rich and powerful. We were also labeled as 'a tiny group of bird-watchers.' But birdwatchers did whatever they could: some wrote letters, some offered ideas, some helped linking up with fellow NGOs and the media, some took legal actions, etc."[62]

The defense of Long Valley was a formative experience for the HKBWS in communicating with the public. Information about the movement appeared several times on the front page of *South China Morning Post*. Experts from the network BirdLife International came to Hong Kong to examine the Environmental Impact Assessment of the KCR project and showed that the conservation value of the wetland had been downplayed. Mike Kilburn, an active member of HKBWS Conservation Committee trained in public relations, wrote an article about Long Valley in the journal of BirdLife International, *World Bird Watch*, in 2000.[63]

This success encouraged the HKBWS to resist the government's precautionary decisions regarding the risks of transmission of avian influenza from wild birds to humans. In March 2004, the government closed the Mai Po reserve because a wild bird infected with H5N1 had been found within a three-kilometer radius of the premises. This decision was highly criticized and rapidly reversed as an excess of precaution.[64] But it was then repeated almost every year, with a twenty-one-day ban imposed on the reserve when infected birds were found in its vicinity. Hong Kong birdwatchers argued that the wild birds found with H5N1 in the territory were resident species and not migratory and that it was therefore irrational to close Mai Po and not bird parks in Kowloon. While urban visitors might be in contact with bird feathers or feces, birdwatchers said, "It is not possible to catch avian influenza through binoculars!"[65] Birdwatchers used ornithology and basic logic to criticize the blaming of wild birds out of fear of avian influenza. They noted that it was easier to close a former colonial institution where a few thousand eco-tourists come every year than the urban spaces in which millions of citizens are in contact with wild birds or live poultry. This political decision denied the knowledge they had built for two centuries as a colonial and postcolonial group of passionate birdwatchers.

In May 2007, Mike Kilburn and Malik Peiris organized a joint conference between the HKBWS and the Department of Microbiology of HKU. They showed the journalists a map where the cases of H5N1 on wild birds were reported. It was clear that most of the cases occurred in Kowloon, and the map was strangely reminiscent of the SARS crisis that had started in the peninsula. The epicenter of the spread of H5N1 from wild birds was the bird market, where Mike Kilburn argued wild birds were illegally traded in a way that the government refused to regulate. He later told me,

> Mai Po is probably the most tested place in the world for wild birds, and no wild birds have been found with the disease. I know that one or two dead birds have been found in the area. But in terms of the claim that it's migratory birds that are spreading death to humans, it's nonsense. My frustration is that so little research has been done in the wild bird trade. But the fact is that birds are an easy way for the government to lay the blame elsewhere. If you say, "shoot the birds," you have to contend with a few green groups. If you say, "close down the poultry farms," you have to contend with the global agricultural industry.[66]

Mike Kilburn thus contrasted the ordinary knowledge of birdwatchers, based on regular monitoring of biodiversity, to the alarms raised by fearmongers based on a few cases. He was proud to mention that the Agriculture Fisheries and Conservation Department had commissioned the HKBWS for a water bird survey of the Mai Po marshes and for a study of egret colonies on the whole territory.[67] The competence that birdwatchers had developed in identifying and counting bird species was used for a general assessment of the biodiversity of the territory, considered as threatened by ecological changes.

> We've been collecting records of birds for fifty years in Hong Kong. This gives us an authority that nobody can question on birds, because the HKBWS was started by English birdwatchers who had this amateur birdwatching model: You write down the birds that you see and you submit your records to the society at the end of the year, and those records are available for anybody who wants to use them. The Conservation department has avowed to me that when they try to catch up an area of biodiversity, they don't see the point to try to compete with the birdwatching society.[68]

If the HKBWS plays the role of a collective sentinel at the level of the territory, one of its members, Geoff Welsh, has taken on a singular position as an individual sentinel. Trained as a birdwatcher during his teenage years in England, he worked as a businessman in Hong Kong. After his retirement, he returned to what he perceived as a hobby and invested in birdwatching the rigor he had learned during his career. He was spending three days a week on Po Toi, an island south of Hong Kong that is inhabited by only a few fishermen, where he counted seabirds. His data and pictures were then posted on the website of the HKBWS and highly appreciated by other members of the association because they gave the best demonstration of the extinction of some resident species and the arrival of new ones. While other birdwatchers practiced monitoring as a hobby, randomly returning to the same places, Geoff Welsh did it systematically, turning a deserted island into a sentinel post. However, he was insistent about not presenting himself as an environmental militant warning about climate change; he contented himself with producing valuable records, leaving the work of interpretation to others. "I love numbers," he told me, "I've done that all my life."[69]

The ability of Hong Kong citizens to build a democratic sentinel under the threat of avian influenza can be compared to the work that Taiwanese birdwatchers were doing at the same time. The political fates of Hong Kong and Taiwan are linked in that China has a pretension to include both in its policy of a reunified "Great Nation," and both have responded by asserting their democratic identities, opening new forms of civil society within the Chinese cultural area. The main difference between them is that Hong Kong, after being a British colony for 150 years, is now under Chinese sovereignty while Taiwan—or the Republic of China (ROC)—is an independent country, although not officially recognized by the majority of sovereign states, and has been under American protection for the first fifty years of its independence. Another important difference is that Hong Kong's minorities are Europeans and Indians, anchoring it in the trade routes between East and West, while Taiwan's minorities are Aboriginals, opening to a politics of "indigenization" that detaches Taiwan from the Chinese continent and anchors it in the Pacific Ocean and its Micronesian diasporas.

How do these political differences influence the constitution of birdwatching societies as sentinels of environmental threats? Being two islands in the Chinese Sea, Hong Kong and Taiwan are crucial sites on the East Asian–Australian flyways and host the same migratory birds. However, due to their differences in size and ecology, they do not have the same resident birds. Taiwan has many more forest birds living in the mountains, which constitute the biggest part of the island. Consequently, there is a whole politics of "endemic species" that differentiates Taiwan from Hong Kong as part of its process of "indigenization" (*bentuhua*). If Hong Kong birdwatchers tend to privilege migratory birds—the perception by a trained birder of a "first-seen" in a familiar habitat—Taiwanese birdwatchers pay more attention to resident birds, with the hope of attracting "eco-tourists" to spot endemic species that have never been seen elsewhere.

While British and Japanese ornithologists collected a lot of specimens of Taiwanese birds in the nineteenth and first half of the twentieth centuries, the development of birdwatching in Taiwan started with Sheldon Severinghaus in the 1970s. A former student of Cornell University who taught French and English at Tunghai University in Taichung, he was in charge of the Migratory Animal Pathological Survey for Taiwan. Through this program, he trained Chinese workers to catch, band, and ring birds. Some of them became prominent birdwatchers, such as Lucia Liu—who married him, studied ornithology at Cornell, and was hired at the Academia Sinica—and Peter Chen, who studied ecology in the United States and later taught it at Tunghai University. While Sheldon Severinghaus published the first field guide on Taiwan's birds in English and Chinese in 1970, Lucia Liu Severinghaus edited the three volumes of *The Avifauna of Taiwan* in 2010.[70] Here is how Lucia Liu and Peter Chen recall the first years of birdwatching in Taiwan:

> The MAPS program was by name focused on pathology. But Dr. McClure was also interested in bird migrations. So the pathological study served as an umbrella, and team members did bird work. But the people who were involved in Taiwan in the early days were not birdwatchers. For some of them it was a job, and the conditions were tough. Some of them did not enjoy fieldwork that much. It was very different from members of birdwatching societies who go out for pleasure now.[71]

> We set up nests to put rings, with two numbers on the ring: an identification number and a mailing box in Hong Kong. We measured the weight, the body length, wing length. We collected parasites for pathological survey, inside and outside. We put dry powder on the plumage and lay the bird on a sheet of paper, and the parasites would fall on the sheet. We cut the central nail of the leg and collected the grease. Then we sent samples to headquarters in Bangkok and Tokyo.[72]

The development of birdwatching in Taiwan thus ran parallel to the democratization of civil society. Since it was forbidden to declare an association under martial law in Taiwan, a "bird club" was set up in Taipei in 1973 for Westerners who wanted to practice birdwatching, with the tutorship of the Animal Protection Association sponsored by Chiang Kai-shek. Similar clubs were created in Taichung in 1975 and Kaohsiung in 1979. During the military regime, wearing binoculars in public was associated with spying activities and was reserved for authorized specialists. The first non-

Western members were Lucia Liu, Peter Chen, and You Hanting (known as Hunter Eu), who had been trained in natural-parks management in the United States and made his career in the National Tourism Bureau of Taiwan.[73] Lucia Liu recalls,

> The first meetings were in the homes of one Westerner, and usually there were a dozen Western people, two or three Chinese from what I can remember, and mostly Westerners. By the end of the 1970s, there were almost no Westerners; it was entirely Chinese. I was in the States for my degree, but my husband came back to work in Taiwan in 1980. When he joined the Taipei bird society outings, he was the only Westerner; the others were all local Chinese. And it was difficult in those days: private cars were rare, binoculars were rare, people gathered and they took buses to wherever they were going, and they shared the few pairs of binoculars. In terms of birdwatching, those were really hard days. But people were really interested, they went every week, and the group just grew.[74]

The lifting of martial law in 1987 opened up not only the possibility to organize birdwatching activities publicly, but also a growing desire to enjoy leisure outside of working hours. The number of societies increased to nineteen and formed a federation, which joined the network BirdLife International in 1996 and currently claims to have 5,000 members. Under the politics of indigenization led by the Democratic Progressive Party president Chen Shui-bian, the Chinese Wild Bird Society became the Wild Bird Federation of Taiwan around 2000; it returned to its previous name after the fall of Chen Shui-bian in 2008. As sociologist H. H. Michael Hsiao observes, "This formerly apolitical organization was gradually drawn into political activism because of the increased destruction of Taiwan's natural environment, which sharply contradicted the Society's fundamental beliefs."[75] While many observers of birdwatching societies in Hong Kong and Taiwan have stressed their democratic organization to contrast it with authoritarian practices of the Guomindang government in Taipei or the Communist regime in Beijing, I don't want to take the notion of democracy as self-evident, since there are also birdwatching groups in China who practice their own form of democracy, and I will later show how the identities of Hong Kong and Taiwan also come from the heritage of colonialism and the Cold War. In this chapter, I want to ask how the participation of ordinary citizens in birdwatching affects the capacity of a sentinel to perceive early warning signals on a frontier, with the risks of being lured by wrong signals.

FIGURE 4.6. The Taipei Bird Club at a raptor festival in Guanyinshan. Photo by Frédéric Keck, April 2013.

The Wild Bird Federation of Taiwan has two major success stories in its rise as a democratic sentinel, the first very positive and the second more ambivalent. In 1996, the creation of the natural reserve of Guandu, after a ten-year campaign against construction projects, was the federation's first major success.[76] This park in the suburbs of Taipei is home to shorebirds and attracts around 10,000 people in October for the federation's annual bird fair. It can be compared to Hong Kong's Mai Po for its role in popularizing birdwatching. Another mobilization presents an interesting contrast with that of Long Valley, which occurred at the same time. The Chinese Wild Bird Society supported the fight that villagers in Huben, Yunlin County, took up against a gravel-extraction project. They argued that the Huben site was home to the Fairy Pitta (*Pitta nympha*), a colorful migratory bird considered endemic in Taiwan. They launched an international petition, "Save the pitta's home. Stop the gravel extraction," and collected more than 10,000 signatures, including 95 percent of the villagers from Huben. The Council of Agriculture carried out a survey of the Fairy Pitta breeding population and estimated a population of forty in Yunlin county. On June

14, 2000, newly elected president Chen Shui-bian declared, "If Taiwan lost the Fairy Pitta, we would not only lose the most beautiful thing in Taiwan, but the whole world would be a poorer place." Consequently, Huben Village was designated as an Important Bird Area (IBA) by BirdLife International and as an ecological village by the ROC government.[77] However, further studies led by the Institute of Endemic Species showed that the Fairy Pitta was spread all over Taiwan and attributed the overall decline of its population to the deforestation of Borneo, where they spend the winter. Consequently, after a long process of environmental impact assessment, the construction of a dam in the village of Huben was authorized in 2008.

The failure of the birdwatchers' mobilization in Huben, which coincides with the ending of the first democratic presidency in Taiwan, may come from the fact that it was organized around a "flagship species"—that is, a species with a high iconic or symbolic value that attracts different actors to get "on board" with an environmental movement.[78] By contrast, the HKBWS did not focus on one single species in Long Valley, but insisted on the ecological value of the habitat managed by local villagers. To sustain the conservation policy in this agricultural land, they asked birdwatchers to regularly visit the area and post the pictures they took of local birds on the society's website. The simultaneous environmental movements in Long Valley and in Huben can be described as two ways to raise an alarm about the fate of birds. While the protection of a flagship species is at risk if a study shows that this species is not threatened as it was believed to be, the conservation of a sentinel territory, on a border where the effects of environmental destruction are visible in the long term, appears more sustainable.

Another contrast between birdwatchers mobilizations in Hong Kong and Taiwan is instructive. There is a territory in Taiwan that occupies a position very similar to that of Long Valley. Kinmen is an island on the shores of the Chinese Sea, very close to Xiamen, that remained a property of the Republic of China and became the site of several fights between the two Chinas. The Wild Bird Society of Kinmen has written a petition against the planned construction of a bridge between Kinmen and Xiamen, which, they argued, would destroy the rich biodiversity of the island, preserved by half a century of military presence. They found the support of the Wild Bird Society of Xiamen, who showed that with the same ecosystem and a higher level of development, many species found in Kinmen have disappeared from Xiamen. But local inhabitants of the island support the construction project, arguing that after half a century of military isolation, the bridge would bring tourists to enjoy the rich cultural heritage (Hakka houses from the seventeenth century and museums of war) as well as products of local vineyards. By contrast with Hong Kong, a military sentinel failed to be turned into an environmental sentinel because the threats were not perceived to be high enough to trigger a local mobilization.[79]

The success of Hong Kong birdwatchers as an environmental sentinel can thus be explained as a middle path between two limits that can be observed in Taiwan: an excessive alert (the mobilization to protect the Fairy Pitta in Huben as an endemic species) and a default of alert (the lack of engagement of Kinmen's inhabitants in conserving their territory). The fate of Hong Kong's birds produced meaningful signals between the symbolic value of a flagship species and the confused perception of threats to a whole environment. This raises a general question for sentinel devices: how to adequately signal a threat in the environment? Can sentinels lure those who watch them, and how can they trust their signals? The answer to that question, which has become crucial for global threats at the scale of the planet, can be found in the fundamental reflections of immunologists at the level of the organism.

SENTINEL CELLS

After broadening the survey of sentinel devices from a farm to a local territory and to a global environment, it is now necessary to go down to the level of the organism, where the discovery of "sentinel cells" has brought immunologists to reflect on the failures and successes of early warning signals. Biologists have come to criticize the view of microbes as enemies and of immunity as a distinction between self and other, leading to a more complex view of the chains of signals in the organism.[80] How does it change the view of sentinels in the immune system?

I first heard about sentinel cells when I followed the work of Jean Millet at the Hong Kong University Pasteur Research Centre, where I was a visiting researcher for the 2008–9 academic year. In this center for research and teaching created in 1999 by the Pasteur Institute in Paris, I regularly followed lab meetings and conferences to learn about the basics of virology, immunology, and cell imaging. But it took me a year to dare ask a researcher if I could watch experiments in the lab, given the strict biosafety rules and the stringent time constraints they imposed. The Pasteur Research Centre had a Biosafety-Level 2 (BSL2) lab that allowed its researchers to work on virus-like particles, studying their entrance and exit from cells; but they had to go to the BSL3 lab in the Department of Microbiology of HKU if they wanted to experiment with live viruses on animals. Both labs worked in collaboration on the viruses of avian influenza, SARS, and dengue, which raised similar problems as pathologies. In contrast, more lethal viruses such as smallpox or Ebola require a BSL4 lab. But the Pasteur Centre specialized in the signals produced in the relations between cells and viruses and didn't need to experiment on live viruses.

Jean was born of a Japanese mother and had been raised at the French high school of Hong Kong. He had just returned from earning a PhD at the Pasteur Institute of Paris and was working on a post-doctorate on coronaviruses, such as those that caused SARS. He held an intermediary position between the Asian and French researchers working in the lab, who had difficulty communicating with each other. While the first group considered emerging infectious diseases to be dangerous threats to the local community and engaged in virology to support local public health, the second viewed these diseases as opportunities to get international funding for research on fundamental mechanisms of virus-cell interaction. Jean could express the political imaginary of Hong Kong citizens in the language of French biological science, which made his guidance valuable.

His experiments were designed to identify which proteins caused the attachment of the SARS virus to the cell it would then invade. Outside the BSL2 lab, he had made double-hybrid tests on yeasts showing that the ezrine protein of the cell interacted with the Spike protein of the virus. "Yeast smells good," he told me, "You feel like you're in a cake factory!" The work on yeast is a cornerstone of Pasteurian microbiology: the Pasteur Research Centre of HKU was first created to map the genome of yeast, before turning to emerging infectious diseases in 2003. Then Jean confirmed his hypotheses in the BSL2 lab by combining virus-like particles (VLP)—the backbone of an HIV retrovirus with Spike proteins from a SARS coronavirus—and monkey kidney epithelial cells (MDCK). These combinations showed the first steps of infection of the cells but remained noninfectious. Jean stored them in a fridge to keep them "fresh."

I had attended lab meetings during which researchers discussed the level of "freshness" that was required for cells to produce meaningful signals of infection. After twenty infections, cells are considered old and must be destroyed because they do not communicate signals well enough to other cells. To maintain freshness, cells are fed a "nourishing milieu" containing calf serum, CO_2, and a liquid called PBS that avoids osmotic shock when they are taken out of their recipient.[81] While watching Jean work in the

BSL2 lab, I thought of the analogies with the model farm of Wang Yichuan where I had worked under high biosecurity conditions; but in the lab, cells had taken the place of chickens as objects of care.[82]

Jean took viral particles from a box and inserted them into twenty-seven wells containing cells and another twenty-seven for mock experiment. "I don't know which well contains cells and which doesn't," he said. "I have to do the gestures in a very repetitive way so that I am not stressed when I come to the virus."[83] His stress did not come from the fear of lethality of the virus but from the will to produce a meaningful result. Jean explained to me that biosafety measures in the lab are aimed less at protecting the researcher from carrying an infection outside the lab than at protecting the experiment from a pathogen that would blur the results.

These results came after a few minutes. First, cells were situated at the bottom of the well, while viruses remained at the top as "supernatant"; then, when the infection occurred, cells converged with each other, and most of them were destroyed. Jean explained this phenomenon as "overcrowding," and for a moment adopted the perspective of infected cells when he said, "Hey girls, we don't have room any more, we'll have to eliminate ourselves!" After this process, some cells remained at the center of the well; they looked "happy," Jean told me, which means that the genetic information of viruses and cells matched.[84] To justify his projection in the cell's interaction with viruses, Jean referred to Barbara McClintock's idea of "feeling for the organism." McClintock has revolutionized the study of cells by looking at the transposition of genes in corn cells.[85] She insisted on the need to "grow" or "breed" cells to understand their genetic mechanisms. As the first virus was observed on a tobacco plant,[86] taking the perspective of a plant cell allowed virologists to see how viruses disrupt organisms.

Jean showed me images on an electronic microscope of infections produced by his colleagues at the BSL3 lab, which confirmed the results he obtained with VLP in the BSL2 lab. Fluorescence revealed the interactions between the proteins of the cells and of viruses, indicating a colocalization. Choosing images randomly, Jean commented that some infections were "beautiful" while others were "ugly." In the second case, cells literally exploded, releasing huge quantities of viruses that jumped to other cells. Jean called it a "dirty death," or necrosis. In the first case, by contrast, other cells spread their arms as if they were calling for help, and viruses were passing from cell to cell through these bridges. This was considered a "clean death," or apoptosis: the information of the virus was communicated to other cells in such a way that they could develop defenses. While the first images looked like a battlefield opposing big cells and small viruses, the second showed a network of cells signaling the passage of viruses.

These cells spreading out their arms were called *dendritic cells* because of their analogies with the dendrones in the synapses of the brain.[87] They were also called *sentinel cells* or *antigen-presenting cells* because they move to the first line of defense of the immune system to capture the information from a pathogen. The immune system is conceived as a set of cells, occupying between 15 and 20 percent of the human body, that communicate with each other to signal a disruption by an unknown biological agent. A distinction has recently been made between innate immunity, which can be found in any living beings and is particularly well known in plants and insects, and adaptive immunity, which exists only for vertebrate animals and opens a space for a biological memory enhanced by vaccination. While innate immunity is produced by cells that recognize specific pathogens, such as the Toll-Like Receptor discovered by Jules Hoffmann in flies and by Bruce Beutler in mice, adaptive immunity is activated by cells that can detect the information from any pathogen, such as the dendritic cells studied by Ralph Steinman in mice and humans. Hoffmann, Beutler, and Steinman received the Nobel Prize in 2011 for these discoveries, which revolutionized immunology.

Philippe Kourilsky, former director of the Pasteur Institute in Paris and professor of molecular immunology at the Collège de France, defines dendritic cells as "professional sentinels."[88] While all the cells of the immune system practice a daily surveillance of the disruptions of the organism, dendritic cells are the only ones that can activate a new immunological response to an unknown pathogen. They have hundreds of sensors that detect specific proteins in the pathogens and receptors that can attach to other immune cells (B cells, T cells, macrophages, natural killer cells). This attachment requires chemical molecules called cytokines and chemokines, which form a cascade of signaling described by pathologists as an "inflammation"—a normal process through which the body regulates abnormal information.[89] If dendritic cells are not the first to detect infection, defined as a higher concentration of chemokines or cytokines than normal, they orchestrate the responses of the whole immune system through interpreting the signals they receive and orienting the responses that follow, thus amplifying the alert.[90]

Kourilsky takes the example of what happens when a human individual touches her eyes after shaking hands with someone who carries the influenza virus. T cells of the innate immune system try to recognize the antigen of the virus, but as the influenza virus is unstable, and as there are "only" one billion specific T cells in the organism, they need to bring the antigen to the dendritic cells of the adaptive system. These cells have the capacity to analyze the information carried by 50,000 T cells and find the adequate receptor for a new virus. They can thus be described, following David Napier, as "search engines for the information (harmful or helpful) that sits latently in viruses."[91] If the information to react to an unknown pathogen is already there in the organism, dendritic cells are the highly mobile synapses that can move around the organism to match it with the information of the pathogen.

This process of searching is crucial for the life of the organism. When cells carry information that they do not recognize, they can undergo apoptosis or try to communicate this information to survive. "Everything is a question of signals in a world of cells dominated by apoptosis," writes Kourilsky. "A cell that does not receive signals of survival has to commit suicide. The normality of cells that constitute life is thus checked by others in a multiplicity of ways."[92] Kourilsky compares dendritic cells to "an army of drones deployed in the organism, endowed with general instructions but capable of adjusting their prescribed goals depending on the environment they encounter." Yet he notes that this army of drones is not commanded by a center but by a field of chemical mediators that constantly detect new events in the living organism.[93]

Following this new view of the immune system as a network of surveillance constantly capturing information, pathologies cannot be explained by the invasion of a foreign agent—it is increasingly recognized that the organism is coevolving symbiotically with the microbes it contains, which makes it difficult to identify a microbe as entirely new—but by a failure of the immune system to adequately activate its cascade of signaling. Many commentators have focused on autoimmunity, a process through which the organism triggers immune defenses against its own tissues.[94] But the research made at the HKU Pasteur Research Centre on emerging infectious diseases points to another direction.

Malik Peiris, who had been appointed scientific director of the Pasteur Centre in 2007, hypothesized that the lethality of avian influenza viruses in humans came from an excess of chemical mediators invading the respiratory track, what he called "cytokine storm."[95] This hypothesis was highly disputed by flu experts, notably Robert Webster, who claimed that it was not confirmed in mice.[96] But it found its principles in the first works led by Peiris and others on the pathology of the dengue virus, explained by a mechanism known as "anti-body dependent enhancement."[97] The dengue virus is more lethal when the organism is infected by two variant strains of the virus because the immune system produces antibodies for the first strain that become excessive for the second strain. The lethality of emerging viruses could thus be linked to a panic of the innate immune system when it is presented with the wrong information.[98]

In this hypothesis, dendritic cells play a major role, since they are supposed to act as modulators of the adaptive immune response. The hypothesis of Peiris and his team is that these sentinel cells are bypassed by lethal viruses such as H5N1 and the SARS coronavirus, which lure or trick them about their antigenic profile. Italian immunologist Alberto Mantovani has shown that some proteins act as "decoy receptors" or "molecular traps" that silence the inflammatory production of cytokines and chemokines. Viruses, following this hypothesis, could mimic these proteins and "scavenge" the inflammatory response—that is, block the production of signals of the receptors that should recognize them and invade the silenced immune cells.[99] Pharmaceutical innovations in vaccinology try to simulate these decoy proteins on computers to artificially regulate the immune system.

The words *decoy* and *scavenge* come from hunting practices. A scavenger is a predator that feeds on dead prey. A decoy mimics the behavior of a predator to lure and attract its prey. In England, "duck decoys" (from the Dutch *endenkooi*, duck-cage) are wooden structures built around a pond; a dog is posted at the end of the pond, and ducks collectively swim toward the dog to inspect what is troubling the water. This trap was built after observing that ducks attack a fox who dares to come into the water, where as a group they have an advantage, or come to inspect if the fox is playing in the weeds showing its tail. A decoy is therefore more than a lure that attracts animals based on a biological desire; a decoy attributes intentions, or at least signaling behaviors, to the animal it is designed to catch. The vocabulary of hunting is often applied to relations between viruses and cells to describe strategies through which viruses avoid the defenses of cells or dissimulate their information to enter the cells. Philippe Kourilsky thus notes, "When I write that a predator feeds on its prey, it doesn't mean that it devours it. Infectious agents are tiny predators that develop (at the expense, for instance, of a human being) by levying energy and distracting for their benefit the biological constituency of their host."[100] Hunting relations are not based on killing and devouring, which are accidental endpoints, but on communicating through signals in a shared environment.

Consequently, when a human being is killed by a virus that it should host, such as an avian influenza virus, it is in fact overreacting to a decoy built by the virus to facilitate its entry into the host. Dendritic cells are smart, but when their behavior is mimicked by the agents they are designed to catch, their intelligence turns into a liability, as it sends the wrong signals of alert to the rest of the immune system. The pathologies of the immune system can thus be analyzed as failures in signaling when sentinel cells are lured by a decoy. This casts light on other levels—political and ecological—where sentinel devices regulate relations between self and non-self. Does this excessive production of signs at the level of cells also characterize sentinels at the other levels, and what does it reveal of the ontology of sentinel devices?

SENTINEL BEHAVIORS

The functions and dysfunctions of sentinel devices at different levels of the life of avian influenza can be compared to the research that has been done in the same period around endocrine disruptors. In that research also, birds carried signs of a catastrophe that threatens humans and displayed a new mode of signaling that develops at different ontological levels.

The alert on endocrine disruptors was raised by Theo Colborn, who followed the footsteps of Rachel Carson in studying the health of wildlife, and won the Rachel Carson Prize in 1999. At the end of the 1980s, Colborn took part in a Canadian project on the impact of pesticides on the health of the Great Lakes. While she did not find cancer, which she expected would explain the decrease in fish numbers, she discovered that seagulls had stopped hatching eggs and that their reproduction had waned. She compared these findings to the experiments of Michael Fry at the University of California, Davis, who had injected DDT into seagull eggs from uncontaminated areas and found "female cell types in the testicles." She also referred to the research of Frederick vom Saal on the endocrine system. He had shown that the disposition of mice in the womb exposed the embryos to an exchange of hormones between males and females; a female placed between two males is more aggressive than if she is placed between two females. If hormones produced by glands (ovaries, thyroid, hypophyse) are carried by blood and send messages to organs on how to develop, then sex depends on environmental exposure in the womb and not only on X or Y chromosomes. Hormones, just like chemical mediators of the immune system, act in a field of signals that they are designed to regulate, which allows biologists to speak of an "endocrine system."

Putting these different data together led Theo Colborn to "move beyond cancer"[101] and propose a new paradigm to explain the effect of chemicals on organisms, which became famous under the Wingspread Declaration in 1991. Chemicals such as DDT, DES, and PCB mimic hormones because their structure is similar.[102] They cheat the endocrine system and disrupt it by attaching to its receptors. Colborn described their behavior as "stealth": they enter the body at a dose below the levels of toxicological perceptibility and are rejected if they enter at higher doses. It was therefore necessary for Colborn to move out of the toxicological paradigm of cancer, according to which "the dose makes the poison," and to invent modes of perception of early warning signals at low doses. The feminized behavior of male seagulls that hatch eggs in the Great Lakes is a "weak noise": even if seagulls do not die of this behavior, it signals a threat to the fate of their species and of the humans who share the same environment.[103] Seagulls, due to their position in the trophic chain as fish predators, are amplifiers of a chemical pollution that affects the whole environment of the area.

In this fascinating and frightening story of biological research, the notion of sentinel—even if it is not used by Colborn herself—could be used to describe four levels of signaling: the interaction between endocrine disruptors and the endocrine system, the "feminized male seagull," the Great Lakes as a political territory between Canada and the United States, North America as an environment for wildlife affected by chemical pollution.[104] At these four levels, the problem for scientists is to perceive adequately the early warning signals that are blurred by the behaviors of chemical industries—who cheat political regulations—and of the agents they produce—who trick biological protections. A form of citizen science is necessary to multiply the information stonewalled by the industry, such as the engagement of birdwatchers in Hong Kong or the formation by Theo Colborn of networks of surveillance for chemical disruptions that seem to cause illnesses among those living close to fracking pads.[105] Endocrine disruptors affect predators—both wildlife species and the scientist who "hunts" for the cause of their decrease—because they act themselves as predators, luring those who try to catch them.

Another research, led by Israeli ornithologist Amotz Zahavi, can be compared to the work of bird flu experts in Hong Kong. He and his wife, Avishag, built a theory of evolution known as the "handicap principle" after observing babblers in the Negev desert for twenty years. As birds became accustomed to the researchers' presence, they could watch their behavior without disturbing them. Some of the babblers, called "sentinels," were sitting on a branch while others were feeding, and they barked noisily when a predator came close. But sometimes the other birds, instead of fleeing, joined the sentinels on the branch to bark even more loudly. "Why do babblers," asked Zahavi, "raise the volume of their calls to a level that can alert a predator to their presence precisely when a predator is in the area—and before it has had a chance to notice the well-camouflaged group?"[106]

When Zahavi calls these babblers "sentinels," he refers to his experience as a soldier in the Israeli army, watching for signs of attack from neighboring countries.[107] But this does not mean that he projected his view of society on bird behavior; rather, he discovered on the border of Israel new means of signaling that led him to recast the foundations of biology.[108] Zahavi compares the behavior of babbler sentinels to the peacock's tail, a classical enigma of Darwin's theory of evolution. Warning about a predator constitutes a handicap in a utilitarian rationality: it presents a risk of being caught, but gives prestige to those who act as sentinels. Sentinels send what Zahavi calls "costly signals": they indicate to the predator that it has been seen and that it would waste time trying to catch its prey, in the same way that the peacock's tail indicates to its rivals in mating that they would lose the fight. These signals establish a communication between predator and prey, or between two sexual competitors, so that both can assess their benefit from an uncertain interaction.[109] Instead of actually fighting with predators or rivals, sentinels threaten them; they send signals of a possible encounter that is uncertain for both parties.

Zahavi's observations are key cases in neo-Darwinian anthropology, which relies on a strong utilitarian view of individuals maximizing their genetic potential; but they push this paradigm to its limit by revealing new forms of communication between individuals. While the behavior of sentinel birds has often been explained as an altruistic sacrifice of the individual to the group, Zahavi asserts that individuals have an egoistic motivation to show their prestige. He does not look at the relation between the individual and the group through sexual selection, but at the relation between individuals through signal communication. "Signal selection differs from sexual selection in that it includes all signals—not just those that affect potential mates and sexual rivals, but also signals sent to all other rivals, partners, enemies, or anybody else. . . . The theory of signal selection thus offers new ways of looking at every species on earth, from microscopic organisms to humankind itself."[110] Just like Malik Peiris had moved away from the view of bird flu viruses as enemies and Theo Colborn from the paradigm of toxic doses in cancerology, Amotz Zahavi refused the description of sentinel birds as sacrificial to analyze the complex chain of signals involved in sentinel behaviors.

It is no surprise, then, if Zahavi took part in a conference on "Immunology as a Cognitive Science" at the Weizmann Institute of Science in 1994, to explain how nonvirulent pathogens become virulent—that is, how they can destroy the cell that is supposed to host them. He considered the relation between pathogen and cell as signal communication, or as an exchange of information about the properties of both partners. "If the host doesn't cooperate," he writes, "it is to the parasite's advantage to act virulently; it is only the hosts' support that enables the non-virulent phenotype to overcome the virulent one."[111] For Zahavi, there is no false signal; signals always transmit information, even when they pass from threat to aggression. Biological virulence or political violence appear as excesses of communication when two predators lure each other.

By contrast, signal communication is for Zahavi a space of aesthetic creation or fiction. While he borrows his models from sexual competition and seduction, which is at the center of neo-Darwinian biology, he orients the description of hunting relationships toward chasing or dancing.[112] If sentinel behaviors oscillate between a default of communication and an excess of communication, what do they communicate about? Zahavi's "theory of costly signals" shows that they communicate about the value of the sentinel.[113] This would explain why sentinel devices use fiction to create the belief that the threat is already there and to produce trust in the capacity of the security system to mitigate this threat. They accumulate signs in a form of ostentatious waste. Zahavi thus summarized his handicap theory in "a very simple idea: waste can make sense, because by wasting one proves conclusively that one has enough assets to waste and more. The investment—the waste itself—is just what makes the advertisement reliable."[114] Following Zahavi, hunters produce a waste of signs to gain trust in their behavior. Signs do not have value in themselves, but only in relation to the future behavior of those who produce them.[115]

My question at the end of the last section of this chapter was: how to account for the fact that sentinels produce excessive signaling at different levels? A correlated question would be: why are birds good operators for sentinel behaviors? The research of Theo Colborn in the Great Lakes and Amotz Zahavi in Israel, in line with that of Malik Peiris in Hong Kong, shows that birds are particularly apt, given their strong position in predatory chains, to send early warning signals at an intermediary level between microbes and humans, between immune cells and a threatened territory.

Sentinel behaviors oscillate between a default and an excess of early warning signals, between complacency and overreaction. I have identified these movements in the different levels at which birds send signs of threats to humans: at the level of a farm, where they are perceived both as a commodity and as a living being requiring care; at the level of a territory, where they can be perceived as symbols of political threat or as early warning signals of pandemics; at the environmental level, where they can be identified as flagship species or as ordinary dwellers of a habitat; at the level of the cell, where avian influenza viruses can disrupt the signaling cascade of the immune system. In these different levels, signs become values when they transform the potentiality of lure into fiction by instituting a common imaginary. Avian reservoirs are the sites in which these signs are perceived by humans to prepare for future threats. In the next chapters, we will see how these signs work in practice and how they produce value.

CHAPTER FIVE

simulations and reverse scenarios

This chapter looks at the way actors in charge of avian influenza simulate the spread of potentially pandemic pathogens from birds to humans. In the last twenty years, techniques of simulation have multiplied in the world of disaster management. The stated objective of simulations is to rehearse a disaster situation to better prepare for it, to do as if the catastrophe had already happened in order to mitigate its effects. Two kinds of simulation can be distinguished: desktop (or table-top) exercises, relying on computer programs and occurring indoors, and real-ground (or full-scale) drills, including actors in an outdoor setting.[1] Both techniques use worst-case scenarios, imagined by disaster-management professionals who have training in fiction writing. These two kinds of simulations can be found in the practices of microbiologists and public health officials in Singapore and Hong Kong, as well as in those of birdwatchers in Taiwan.

My general argument is that techniques of simulation are transitory forms between the "hunting" practices of microbiologists and birdwatchers when they communicate with animals through sentinel devices, taking the perspective of pathogens in animal reservoirs, and the more "pastoral" techniques of public health officials who plan for the mitigation of pandemic threat at the level of the population, particularly through stockpiling vaccines. Simulating bird diseases, I will argue, is not only a concern for public health but also an environmental practice: it prepares not only for a pandemic disaster but, more fundamentally, for an uncertain encounter between humans and animals, or between predators and prey.[2]

While the genealogy of techniques of simulation usually starts with the debates about nuclear disasters during the Cold War, my analysis discusses concepts of simulation, performance, ritual, and play from the perspective of hunting societies. The perspective of animals in disaster simulations has not been taken by social scientists because these simulations have been considered pastoral techniques of biopower, therefore bypassing what the anthropology of hunting societies shows about ritual action. Virus hunters and birdwatchers simulate the encounter between humans and birds as potentially catastrophic to instantiate relations between species. The perspective of sentinels developed in chapter 4 will be adopted here for simulations that may appear as belonging to a more pastoralist tradition of protecting the population against an enemy. I will ask what it means to imagine a future pandemic by building scenarios involving humans, animals, and artifacts.

To analyse simulations of bird diseases, I combine the notion of reverse scenario with the notion of reflexive ritual. Pandemic or extinction scenarios, I argue, connect humans and animals in realistic ways to play on the imagination of what could happen if a pathogen emerges. Using animal reservoirs as mirrors where anxieties about pandemics are reflected, humans exchange their perspectives with animals in an imagined future where their relations are reversed. Simulations can thus be situated in an anthropological space between play and ritual.

DESKTOP EXERCISE AND REVERSE SCENARIOS

In July 2013, I was visiting Gavin Smith, who was then an associate professor at the Duke Graduate Medical School of the National University of Singapore. I had known Gavin in Hong Kong while he was working at the State Key Laboratory of Emerging Infectious Diseases created by Guan Yi at the Department of Microbiology of the University of Hong Kong. Born in Australia, Gavin had studied botany and ecology at the University of Melbourne and then moved to Hong Kong for a PhD in molecular systematics under the supervision of Guan Yi. Gavin had first studied the genome of fungi, then of the SARS coronavirus,[3] then of avian influenza viruses. He described this development of his research as a reduction of the size of the living beings he was sequencing and a refinement of the data and results he could produce. Only in his forties, Gavin was one of the most renowned experts among flu scientists, benefiting from the support of Robert Webster, with whom he had signed articles in *Science* and *Nature*. Always joyful and good-humored, speaking quickly but taking time to discuss with his interlocutors, he was also politically engaged in the defense of the environment and saw bird flu as an opportunity to fight against the poultry industry or the pollution of the atmosphere.

Gavin had left Hong Kong in 2012 to settle with his family in Singapore. He was followed by his team partner, Vijaykrishna Dhanasekaran (usually called Vijay), hired as an assistant professor at NUS–Duke Graduate Medical School. The last member of his Hong Kong team, Justin Bahl, had left for the University of Texas, but he was an adjunct faculty at NUS–Duke Graduate Medical School, and the three of them continued to publish together. In Hong Kong, they had developed fundamental research on the molecular evolution of influenza viruses, relying strongly on the collaboration between the State Key Laboratory of Emerging Infectious Diseases and the University of Shantou. This university had been created in 1990 by the Hong Kong tycoon Li Ka-Shing in his native Guangdong, with the approval and administrative support of Chinese authorities. It provided the State Key Laboratory of Emerging Infectious Diseases with samples collected in poultry markets in the Fujian province, thus offering a real-time surveillance of influenza strains emerging in this "epicenter of pandemic flu." But after the publications of Gavin and his team showed the role of the Chinese poultry industry in the emergence of these new strains,[4] the Guangdong authorities severely restricted Hong Kong virologists' access to these samples. Much to the dismay of his mentor Guan Yi, Gavin then moved to Singapore, where he had more freedom to continue his fundamental research and more funding from international institutions. He had shifted from the sentinel, where he could track the mutations of flu viruses at the gateway of the epicenter, to a place where he could simulate viral evolution on high-tech computers.

Singapore was then considered the new "pole of excellence" for biological research. The small city-state founded by Lee-Kuan Yew in 1965 at the end of the Malay Peninsula had converted its industrial and financial power, due to its position on the commercial trade routes between East and West, into an economy of knowledge and information, with high investment in biotechnologies. At the turn of the century, the Singapore government launched Biopolis, a biomedical hub combining researchers from private and public sectors, attracting academics from all over the world through financial and technical facilities. Within its original seven buildings connected by sky bridges, it housed five research institutes. One of them, the Genome Institute of Singapore, was headed by Edison Liu. Born in Hong Kong, Edison Liu made his career in the United States before coming to Singapore to run the Genome Institute. He was also the president of the Human Genome Initiative between 2008 and 2013 and succeeded in moving the office of this international consortium from London to Singapore in 2014. He became one of the "scientist heroes" whose figures gained prominence in the media after the SARS crisis, during which Singapore ran a race with Hong Kong and Atlanta to be the first to sequence the new coronavirus, and almost won.[5]

However, Gavin Smith was not working in Biopolis, which is situated close to the main campus of the National University of Singapore in the eastern part of the city, but next to the Ministry of Health and the Singapore General Hospital in the center of the city. The NUS–Duke Graduate Medical School had been created in 2005 to pattern the training of high-tech medical leaders on that of the Duke University School of Medicine in North Carolina, which could veto any decision made by the governing board of the National University of Singapore. The Program in Emerging Infectious Diseases which had hired Gavin was headed by Wang Linfa, who was also leading a team on bat viruses at the Australian Animal Health Laboratory in Geelong.[6] Born in Shanghai, trained at the University of California at Davis, Wang Linfa was editor in chief of the journal *Virology*, where Gavin Smith was associate editor. His program was running projects for the surveillance of emerging infectious diseases, particularly influenza and other respiratory diseases, in Singapore and the rest of Southeast Asian countries.

Gavin's career took a meaningful turn when his team published two major articles at the height of the 2009 H1N1 pandemic. The first one, entitled "Dating the Emergence of Pandemic Influenza Viruses," was published in the *Proceedings of the National Academy of Sciences* USA just after the emergence of the H1N1 virus in Mexico in April 2009, but it had been written before and relied on work accumulated in the previous years. It showed that genetic components of the 1918 H1N1 pandemic virus circulated among pigs and humans as early as 1911, thus challenging the commonly held view that the virus causing Spanish flu jumped from birds to humans in 1918. In the context of a new H1N1 "swine flu" virus, this article stressed the need for an intensified surveillance of pigs to detect emerging flu viruses before they become pandemic. "If future pandemics arise in this manner," Gavin and his team concluded, "this interval may provide the best opportunity for health authorities to intervene to mitigate the effects of a pandemic or even to abort its emergence."[7]

This conclusion was immediately enforced by Gavin's team's second "buzz-article," entitled "Origins and Evolutionary Genomics of the 2009 Swine-Origin H1N1 Influenza A Epidemic," published in *Nature* at the end of 2009, with an online Wiki-version circulating as early as July 2009. In this article, they showed that a "twin" virus bearing the same genetic components as the pandemic H1N1 had been identified in pigs in Hong Kong in 2004. This article was quoted by the Food and Drug Administration in the United States as proof that the pandemic H1N1 virus did not originate in North America but in Asia.[8] Gavin and his team made it clear that their results should not lead to the conclusion that the Asian swine industry was to blame, but on the contrary, that there was a better surveillance of the animal reservoir for influenza viruses (waterfowl and pigs) in Hong Kong than in the United States. Their article ended, "Despite widespread influenza surveillance in humans, the lack of systematic swine surveillance allowed for the undetected persistence and evolution of this potentially pandemic strain for many years."[9]

In July 2009, I took a course at the Hong Kong University–Pasteur Research Centre in bioinformatics, during which Gavin, Vijay, and Justin explained to virology students how to obtain this kind of result. Bioinformatics is a method to treat the massive amount of biological data available on the web—particularly through the website of the U.S. National Center for Biotechnology Information, GenBank—using dedicated softwares. These softwares, called BLAST (Basic Local Alignment Search Tool) and MSA (Multiple Sequence Analysis), calculate the probability that genetic sequences derive from one another. Such a procedure, called alignment, aims at approximating the real descent relationships between viruses and at drawing the optimal "tree of life." It is based on the Darwinian hypothesis that life expands in a rational manner to maximize fitness capacities. Adrian Mackenzie, scholar in Science and Technology Studies, thus explains, "The real problem for bioinformatics becomes calculating the 'optimum alignment,' the minimum number of edits needed to move from one sequence to another. Bioinformatics almost holds as a founding axiom that optimum alignment expresses similarity or kinship between biochemical entities."[10]

Gavin and his team made it clear, however, that bioinformatics is not only about probability calculation, but also involves a work of virtual imagination based on actual experience. In the construction of a phylogenetic tree, biological knowledge is necessary, because some correlations proposed by the computer make no sense in an evolutionary perspective, since they can be due to errors in sequencing or genetic deletion. To decide which correlations to take into account and which to consider irrelevant, virologists use other software (such as Bootstrap, Jukes Cantor, or Tamura) calculating the probability of correlations based on a given scenario. Vijay explained, "Imagine you've sequenced a virus, you want to know the evolution, where it comes from. You download all these sequences from GenBank and then you make the alignment to check which nucleotides are important. But if there's something obscure in the sequence, you just check the references and ask the labs who published the sequence: 'what month did you do the analysis?'"[11] The logical correlations between sequences on the computer needed to be confirmed by chronological derivations between strains.

Gavin thus opened the bioinformatics course with a simulation of the work they had done a few months before on the H1N1 virus, presented as a desktop exercise of what to do when a new virus emerges. "Imagine we've received this new sequence from Atlanta. Paste it to Blast, and take as much information as possible. Make a tree so big you can never print it on a sheet of paper: it will be good starting material before reducing the inquiry."[12] Gavin and his team explained that in contrast to other virologists working on flu viruses, they had decided to modelize the eight segments of the flu RNA, not just the two segments that shape the H and N proteins usually considered to be the most significant markers of antigenic drift. They could thus build eight parallel evolutionary trees that, by their similarities, gave an indication of the real evolution of the whole flu virus. As Justin later said to me, it was better for them to start with a huge amount of data and reduce it progressively to achieve a result that would be biologically meaningful. "Every time we reduce the information, we need to make sure it is consistent. We proceed following a Markov chain where every step forward is controlled by a return on the information of the previous steps. We try to get close to the real tree, which explains genetic sequences: what we call the absolute evolutionary tree."[13]

The images Gavin and his team produced were impressive. By drawing back the "molecular clock" based on the estimated kinship relations between viruses, they allowed virologists to see the branches where evolutionary ruptures occur and pathogens jump from one species to another.

By following continuities between genetic sequences, they exhibited the discontinuities between evolutionary niches. Justin said that when a virus jumps from one species to another, it inaugurates a new set of lines, as the "immunity pressure" of the environment creates an "evolutionary bottleneck" in which this mutation can replicate. Drawing back these lines allows virologists to reach what they assume to be "The Most Recent Common Ancestor." "I love those trees," he added. The amount of information provided in a single image was correlated to the unity of an evolutionary hypothesis drawing this information back to a common ancestor.[14] The image provides an idea of where the discontinuous shifts occur in a continuous game of mutations.

Historians of science and art have shown the role of evolutionary trees in the new visual culture inaugurated by Darwinism.[15] Even if the view of natural competition on which this model relied could be contested by new ecological approaches, I was mostly intrigued by the way Gavin and his team turned figures into images through an anticipatory mode of reasoning that can be called "reverse scenario." Natasha Schüll proposes the term in her work with online gamblers to describe the way they "cope with the necessarily uncertain futures of any hand by returning them to a point in the past and confronting them with a branching diversity of outcomes that might have emerged from it."[16] It applies well to "reverse genetics," a technique used to reconstruct viruses from the past, allowing "virus hunters" to go back in time and imagine potential futures for the evolution of pathogens.[17] Just as a gambler thinks how much he could have won if he had played this or that card, Gavin and his team imagined what could have happened if the H1N1 virus had been detected in pigs prior to 1918 or 2009. "What we are trying to do is to learn from past events and currently circulating viruses to then try and understand what the possible scenarios are if something new emerges," Gavin explained on his homepage.[18]

Through computer simulation, thus, virologists see the life of viruses as if it could be managed on a screen as potential risks. Similar to traders and hunters, they have to imagine the potential movements of pathogens in animal reservoirs to anticipate them as fast as possible, using bioinformatic software as traps or decoys. In that perspective, the risks of pandemic mutations of flu viruses become similar to the risks of financial products that traders, who can be considered sophisticated gamblers equipped with computers, follow on their screens. In his ethnography of a trading office in France, Vincent Lépinay shows that with the advent of global financial derivatives, "the locus of uncertainty has been displaced: it was once the market of one security with its chess player, but it is now the portfolio with its intricate correlation spun by the formula." To explain what a portfolio is, Lépinay uses a metaphor of wildness borrowed by traders themselves. "The unpredictable animation of these portfolios is akin to dealing with wild organisms, and, unsurprisingly, the term 'beast' comes frequently to the forefront in the traders' conversations about their products."[19] Traders are often described through metaphors of contagion, infection, and toxicity as playing on their computers with risky financial products and simulating financial crises to increase their value, but they can also be compared to virologists anticipating the next pandemic from the surveillance of viral mutations in animal reservoirs, whose circulation is analogous to that of portfolios in global markets.[20] Similarly, the almost childish joy of Gavin's team comes from their biological dream of managing evolutionary ruptures and playing online with the potentially pandemic pathogens they create. If simulation is a work of imagination rather than a calculation of probability, it appears as a rational play with the "wild beasts" created by nature and transformed on the screen into genetic sequences.

I now turn from desktop simulations to real-ground exercises, thus moving from virtual imagination to actual performances, and introducing the concerns of public health planners. I will show that the form of the "reverse scenario" is shared by these two types of simulations, combining fiction and reality in an intriguing way.

REAL-GROUND EXERCISES AND REALISTIC SCENARIOS

After my visit to Gavin Smith at NUS–Duke Graduate Medical School, I went to a nearby building to meet Jeffery Cutter at the Singapore Ministry of Health. Jeffery Cutter was an epidemiologist and the director of the Communicable Diseases division of the Ministry of Health. He was also in charge of the coordination of exercises for preparedness for emerging infectious diseases in Singapore. "There are 50 million passengers every year at Singapore Changi Airport," he told me, "so there is a high probability that emerging infectious diseases will come to Singapore. We need to be prepared." During the SARS crisis, he said, 238 people were infected in Singapore with the new coronavirus between March and May, with thirty-three deaths. They were all treated at the Tan Tock Seng Hospital, which had been designated by the government as an isolation hospital for SARS patients, and where three nurses died of the disease. At the peak of the crisis, all schools in the country were closed for ten days, and home quarantine was established for more than 6,000 people who had come into close contact with a SARS-infected person.[21] Singapore's prime minister at the time, Goh Chok Tong, said publicly that SARS could possibly become the worst crisis Singapore had faced since independence. When a journalist asked him if he was being alarmist, Goh responded, "Well, I think I'm being realistic because we do not quite know how this will develop. This is a global problem, and we are at the early stage of the disease."[22] Ten years later, the Ministry of Health was preparing for the arrival of another coronavirus, MERS-CoV, which emerged in early 2012 in Saudi Arabia and spread worldwide, infecting fifty people and killing half of them by July 2013.

Jeffery Cutter introduced me to Yoong-Cheong Wong, head of the Emergency Preparedness and Response Division at the Ministry of Health. Mr. Wong was a former policeman in charge of organizing exercises for emerging infectious diseases in hospitals. The first national exercise for influenza in Singapore was organized in 2006, after the first cases of human-to-human transmission of H5N1 were suspected in Thailand. Called "Sparrowhawk," it was organized in two stages: a table-top discussion and assessment of pandemic response plans among six hospitals in the territory between April and June and a practical exercise on July 21–22, in which the six designated hospitals had to manage potential patients and share information about their circulation. Mr. Wong explained that a thousand people were involved in the real-ground drill, mostly volunteers from grassroots organizations and nurses from training institutions. While patients moved from one hospital to another, some hospitals assessed the behavior of others and gradually raised the number of casualties. "If there are 200 casualties in one hospital, which resources do they need?" That was the main question, according to Mr. Wong. The aim of the exercise was thus to produce knowledge on the management of mass casualties among the medical staff. "Nursing," he told me, "is not a cheap labor. It needs training for a day-to-day readiness."

The goal of these exercises is often presented by public health experts as the identification of gaps in coordination between the actors who manage epidemics. But the actual criticism of planners themselves, who are in charge of writing scenarios, most often focuses on the lack of realism of the drills, and hence the lack of engagement for the actors of the exercise. "We try our best to do it as real as possible," said Mr. Wong. But he recognized that everyone knew "it was just an exercise." Realism was introduced by varying the conditions of the exercise or simply by not saying that it was an exercise, which aimed at producing surprise. To explain the improvement between Sparrowhawk exercises in 2006 and 2008 in Singapore, an official leaflet notes, "The exercises were activated unannounced to inject greater realism. The Ministry of Health Emergency Medical Staff turned up without prior justification at the general ward of the exercise hospital and 'triggered' the simulated case of human infection with avian influenza."[23]

What does it mean for a simulation to be realistic? Here the notion of "reverse scenario" is useful. Just as microbiologists imagine what could have happened if viruses had been stopped in their animal reservoir, public health planners imagine what would happen if contagious patients were stopped before entering hospitals. But in both cases, death is left aside as a dirty reality that cannot enter the form of the scenario; it is given a "reverse" form through the play on data. By contrast, another type of exercise conducted in Singapore was very real. On July 17, 2013, while I was conducting research at the Ministry of Health, the Agriculture and Veterinary Authority (AVA) organized a drill in one of the poultry slaughterhouses on the border with Malaysia. Malay chickens constitute half of the poultry consumption in Singapore, the other half being chilled or frozen chickens imported from Thailand, Brazil, and North America. There are no authorized poultry farms on the territory and only five chicken farms for eggs, which constrains Singapore to import and slaughter from Malaysia forty million live chickens every year. Malaysian farmers need to be licensed by the AVA to import their poultry to Singapore and are always regarded with suspicion by border authorities. This tendency has increased with the threat of bird flu, and yet no case of H5N1 has been found so far in Singapore.

I was able to watch the drill online because it was streamed widely on a number of websites, and I later conducted interviews with the administration who organized it. The name, "Exercise Gallus VII," referred to the fact that it was the seventh bird flu exercise conducted by the AVA since 2002. It took place in the biggest of the ten slaughterhouses in Singapore run by Soonly Food Processing Industry, which processed 80,000 chickens daily, providing 25 percent of the fresh chicken meat in Singapore.[24] The scenario for the exercise was the following: 1,500 chickens had been infected with H5N1 and should be killed and destroyed, according to emergency response plans. About 100 employees of the slaughterhouse, mostly Indians, had to be screened for their temperature, take an antiviral drug, and then wear personal protective equipment (boots, cap, mask, glasses, gloves) on the back of which their name was written. The "affected" chickens arrived in green cages on belts, were immersed in water, and then electrocuted. Dead chickens were then placed inside two layers of biohazard bags to be incinerated. Dr. Yap Him Hoo, director of the Quarantine and Inspection Group, declared on television, "To make sure that AVA is prepared for any incursion, the main thing we want to do for this exercise is to train our officers, as well as the other parties that are involved in the contingency planning, to be familiar with the steps of the procedures."[25] Desmond Tan, food safety officer at the AVA, said for a previous exercise, "After all the practice we had today, we are able to handle culling as well as unloading of chickens from the truck at a much faster pace. I believe the cullers are able to proceed at a faster rate to cull all the chickens."[26]

I will come back to the paternalism and the ethnic, class, and age divides that structure these exercises. But first I want to focus on the division between humans and animals that seems to overdetermine other oppositions. While the simulation at the clinic was animated by a sense of urgency and scarcity of resources confronted with a fake pandemic, the simulation at the slaughterhouse was quietly following the procedures in actually killing false positive chickens. The chickens didn't have to fake illness; they were considered mere commodities to destroy as if they were contaminated. But the workers acted as if the chickens were sick when they took drugs and protective equipment. I was also intrigued by the fact that the Hong Kong government organized many such exercises on the human side, but never simulated bird flu in poultry farms, because outbreaks occurred regularly in poultry farms on its territory. While Singapore authorities actually killed poultry for a disease that remained virtual for its population, Hong Kong authorities simulated the transmission of avian influenza from chickens to humans because it was actually circulating in its bird population. This is what makes "reverse scenario" a connecting form or a scheme for action: if death cannot be realistically represented in the scenario, it is the final event whose anticipation connects all the actors on the stage of the simulation. The form of the reverse scenario was shared by public health planners in Singapore and Hong Kong, as it came from the world of computer gaming; but it was actualized differently depending on how pathogens crossed the species barriers and revealed various relations between humans and animals. Therefore, I use *reverse scenario* as a term to encompass the varying modalities of bird flu scenarios and how they play with the death of animals to anticipate the death of humans.

To foster this analysis, I will now compare exercises for pandemic preparedness in Hong Kong with those of Singapore. Hong Kong's coordination for this kind of exercise for China is very similar to the leading role Singapore played in Southeast Asia. On June 7–8, 2006, Singapore and Australia organized a desktop exercise for the twenty-one members of the Asia-Pacific Economic Cooperation. The Exercise Coordination Centre was based in Canberra, from which information was sent to the twenty-one members of the public health staff about how the scenario would unfold. An outbreak of human-to-human H5N1 influenza was reported in Singapore (called "Straits Flu"); WHO had upgraded the alert from phase 4 to phase 5; a warehouse that manufactures personal protection equipment had burned; teams were flying to Vietnam for an APEC football tournament; backpackers and fishermen started to get sick around the area of the outbreak. On August 14–15, Singapore hosted a post-exercise workshop in which participants agreed, "The exercise provided an excellent opportunity to establish and test a communication network and to develop relationships between the economies."[27]

Similarly, on November 13, 2006, Hong Kong coordinated a desktop exercise called "Great Wall" in which the health departments of Hong Kong, Macau, Jiangsu, and Beijing had to manage a virtual cluster of three human infections with H5N1 in a family after their visit to a poultry market. As one of the members of the family was supposed to travel from Jiangsu to Hong Kong and Macau, the goal of the exercise was "to synchronize the three systems (of pandemic response) to engender an effective response among them."[28] The exercise opened with a ceremony attended by the vice-minister of the Chinese Ministry of Health, Wang Longde. It was considered a success in creating good relations between mainland China and the former British and Portuguese colonies. Simulations thus instantiate continuities after the discontinuities of public health crises: while the SARS crisis had separated China from former colonies in 2003 due to Beijing's reluctance to share information, "Great Wall" unified the country against an imagined enemy, pandemic avian influenza. But it created other ruptures between humans and nonhumans.

In January 2009, after long discussions with the Centre for Health Protection, I was allowed to observe an exercise organized in a Hong Kong hospital. The Centre for Health Protection was created by the Department of Health after the SARS crisis to anticipate similar outbreaks through active surveillance and communication. Within this center, the Emergency Response Branch, headed by a police officer, was in charge of writing scenarios and organizing simulations. Twice a year, a field exercise was held in Hong Kong, with observers from China and overseas. These field exercises bore names of natural phenomena—Maple, Cypress, Chestnut, Redwood, Eagle, Mountain Hua—as if to show that diseases were part of the natural ecosystem of the territory. There was no other epidemic simulation in Hong Kong in 2009 because the Emergency Response Branch considered the management of the H1N1 virus in the spring and autumn to be a "real-life exercise."

"Exercise Redwood" took place in the clinic of Shau Kei Wan, located in the working-class district of Hong Kong island. The following scenario was distributed to the participants and posted at major public buildings. Confirmed cases of avian influenza with human infection had been reported in Hong Kong's neighboring countries, as well as a rising trend in patient attendance with influenza-like illness in Hong Kong hospitals. Live poultry tested positive for influenza in Hong Kong's markets and were then culled at farms and markets. A member of the staff participating in the culling was reported to have the H5N1 virus, as well as an eight-year-old boy who had played with live poultry. Four clinics were designated in Hong Kong to triage patients with influenza-like illness and send patients with H5N1 to emergency departments. Only the final part of the scenario was performed in Shau Kei Wan; the first part was meant to provide a plausible context.

The official purpose of the simulation was to coordinate hospital services in the management of patients with influenza-like illness. Eighty actors playing patients came in through the front door of the hospital and were sent to different departments depending on their symptoms (pulmonary conditions, tuberculosis, etc.). Twenty "players" treated them in the services, and two "simulators" communicated with other hospitals on a hotline. Those diagnosed with H5N1 were evacuated by ambulance through the back door of the hospital, where journalists took pictures. The head of the Department of Health gave a press conference after visiting the hospital, but he received questions about new bacteria found in a private jacuzzi, not about the success of the exercise. The drill was deemed successful by journalists and public health authorities because it had been held. The scenario was designed in such a way that no surprising event could happen: the only reality that came up was the bacteria in the jacuzzi, not the virus in bird markets.

All the actors I saw looked young and relaxed, wearing blue caps and casual clothes. They wore tags around their necks indicating their symptoms, as well as their name, address, nationality, sex, and age. These cards were red or green to indicate whether they had the flu or only influenza-like

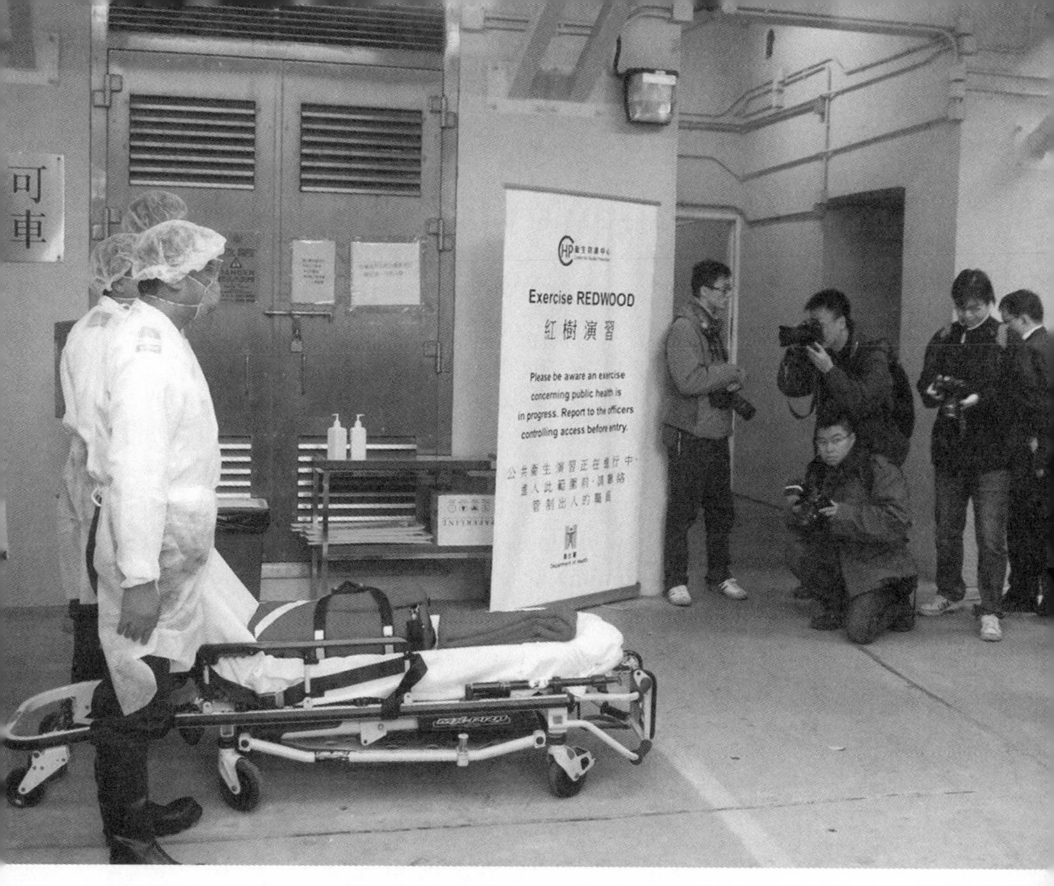

FIGURE 5.1. Exercise Redwood in Shau Kei Wan clinic. Photo by Frédéric Keck, December 2009.

illness. They did not have to fake being sick; their only role was to move to the designated departments. Similarly, in another exercise organized by the Centre for Health Protection, a plane was evacuated because a patient had been found with influenza-like illness; those who sat next to the patient received a red tag, and those who sat at a good distance away received a yellow tag. What was remarkable in Exercise Redwood was the absence of tension or anxiety in the medical staff's behavior. Although humanitarian NGOs consider triage to be an ethical issue when a high number of patients is met with limited resources,[29] the medical staff of Shau Kei Wan clinic only separated symptomatic and asymptomatic patients to avoid the spread of the disease; they did not have to classify them according to their chance of survival in an overwhelmed medical environment. The tension of the situation—who would be considered a spreader, who would receive treatment first—was delegated to artifacts: tags and caps. People were reduced to bearers of signs circulating fluidly in the hospital under the gaze of the media.

While the medical staff of the hospital played their own role, the actors playing patients with influenza-like illness came from a humanitarian association, the Auxiliary Medical Service (AMS). This association, created in 1950 to deal with the influx of Chinese refugees, was registered in 1983 under the Security Department of the Hong Kong government. Its 4,000 members are trained for disaster management in Hong Kong and outside. One of them described it to me as "after-work entertainment," "a place to meet other singles";[30] it is an elite association where similar people express and share moral values. During their own exercises, performed every month, they train how to rescue victims of car accidents or fires. They simulate heartbeats on dummies with fake hearts. "Patients cannot fake the heart rate or the respiratory rhythm," the organizer told me, "so they have tags." Tags have an indirect effect of reality as bearers of signs, while dummies can actually simulate the heartbeat of an injured victim.

It can be said, consequently, that when they simulate patients wearing tags and caps in an epidemic exercise, members of the AMS literally act as if they were dummies. From their perspective, there is a sharp difference between regular accident rescue exercises, in which they take the initiative, and extraordinary epidemic evacuation drills, in which they coordinate with several governmental departments. Members of the AMS I met described their engagement in their own exercises as significant moments in the life of the association and, by contrast, felt frustrated by Exercise Redwood. One of the actors said he felt "passive," playing on the ambivalence of the term: "As we had experienced SARS in 2003, we all can forecast how we would behave in another similar situation, like avian flu. Being a citizen, we are quite passive; I believe that we should follow the guidelines and the Government's advice so as to prevent ourselves from getting affected. As in the exercise, we were being passive."[31] This quote ambivalently says that citizens are passive when they do not know how to behave in an epidemic situation and that actors who follow the scenario of the exercise are also passive because they have no initiative.[32]

There is a fundamental distinction in real-ground exercises between actors and players. While actors are passive, reduced to the artifacts that "act" or "speak for them," simulators are active, since they can introduce uncertainty by the way they combine these artifacts in the scenario. If actors have to appear "real" to produce good images in the media, simulators introduce variations in the situations proposed by the scenario.[33] The distinction between simulators and actors in real-ground simulations runs parallel to the distinction between virologists and public health planners in desktop exercises: while actors and planners only follow their roles in the scenario, simulators and virologists explore the potentialities of "reverse scenarios." It should be noted that simulators were mostly elderly males while the actors were more frequently young and female—a fact to which I will come back in the rest of my analysis.

For simulators, the distinction between real-ground exercise and desktop simulation is not significant, as both combine the virtual and the actual in different ways. In an exercise called "Jadeite," members of the Emergency Response Branch were simulating the evacuation of a residential building. The scenario specified that they had to evacuate the simulation room to a "fallback room." The simulation room, from which the vulnerability of vital infrastructures in Hong Kong was assessed virtually, thus came to be seen itself as a material environment under threat. The head of the Emergency Response Branch who described this exercise to me said it was "more fun than a movie," which evoked a film whose images suddenly burst through the screen. In this reversal of the scenario, simulators became actors and fiction reality. Neither a ritual performance nor a scientific model of prediction, the simulation was for this former policeman a game he enjoyed playing.

The paradoxes of "passive actors" also concern the distinction between humans and nonhumans. Animals can be described as "passive actors" in a sense that is not counterintuitive. If there was no need to simulate the culling of poultry because it regularly happened in Hong Kong markets, simulations for the evacuation of residential buildings in Hong Kong planned that some residents were accompanied by pets that should be handled with care by the Agriculture Department, because their reaction to the evacuation was unknown. Similarly, in simulations of the evacuation of planes, actors were explained how to handle "reluctant potential patients." The real opposition in simulations is thus not between humans and animals but between poultry, considered as commodities that can be destroyed, ordinary citizens, and pets, considered as "passive actors" that should be handled with care, and simulators, who can change the rules of the game. More than sharp ontological distinctions, there is a gradient of activity in the interplay between virtual and real produced by simulation.

The extraordinary context of the simulation thus provides an opportunity to display a series of apparent contradictions: humans become similar to artifacts or animals, actors turn out to be passive, fiction enters into reality. I thus propose a definition of reverse scenario that is different from that proposed by Natasha Schüll: what makes the scenario meaningful in a simulation is not only its capacity to reverse from the present to the past, but also its capacity to change the status of the entities engaged. This capacity to displace contradictions in an extraordinary context through reverse scenarios brings simulations close to rituals, even if it is borrowed from games. Simulations can be considered public health rituals because they regularly justify for the public the work of virus hunters who sound the alarm about future pandemics. However, the analysis of Exercise Redwood incites me to go beyond the simple observation that simulation creates regular order after an extraordinary disorder and to pay attention to how actors are transformed in this extraordinary frame of interaction. Birds, I argue, change meaning when they enter the virtual space of simulation initiated by virus hunters. There is something specific in simulations of avian influenza, by contrast with other epidemics, in that birds act as sentinels of future catastrophes affecting humans. To analyze simulations of avian diseases as rituals of virus hunters, I will now turn again to birdwatchers, thus moving from domestic poultry to wild birds.

BIRDWATCHING EXERCISES AND SCENARIOS OF EXTINCTION

Two types of collective exercises can be distinguished among the birdwatchers I worked with: one involves counting wild birds, the other caring for them. I have observed these exercises in Hong Kong and Taiwan, which, as I have argued in the previous chapters, have invented new practices with wild birds by adapting to the Chinese context models borrowed from Britain and the United States.

Birdwatchers regularly organize bird races, usually in the spring. Coming from traditions of hunters competing with each other for the game, these races are modeled on the Christmas Bird Count, launched by Frank Chapman in 1900 in the U.S., but soon abandoned any reference to a religious marker.[34] They are now animated by the ideal of citizen science, producing knowledge about species extinction out of local engagement.[35] Over the course of one or two days, birdwatchers compete throughout a large territory to spot as many birds as possible. In April 2013, I followed a bird race organized by the Wild Bird Federation of Taiwan at Dasyueshan, a national park in the mountains at the center of the island. For two days, one hundred participants could drive on a 50 km road at 2,000 meters in elevation. The most successful teams saw a hundred bird species out of the 200 that have been identified in this area—a performance on a very wet and foggy day. The race was followed by speeches and gifts, such as binoculars, cameras, books, and tea.

Before the race, participants played a game for which they were divided into three different roles: insects, birds, and trees. They received cards and a scenario in which they had to devour each other to reach an equilibrium among these three components of the forest ecosystem. In an annual festival organized by the Wild Bird Society in Taipei, actors were dressing like birds and dancing in front of the association's leaders with an eagle's giant picture on their back. These bird races and birdwatching festivals can clearly be described as rituals, since they create extraordinary sets of space and time during which humans come closer to birds and imagine that they "are" birds. Contrary to simulations of avian influenza pandemics, they don't rely on the imagination of a pathogen passing from birds to humans, but on the potential extinction of a bird species, perceived as a signal of the potential extinction of the human species. These extraordinary simulations are based on the ordinary work of databases: viral sequences on GenBank in the case of virus hunters, the red list of endangered species of IUCN in the case of birdwatchers.

Coming down from the bird race in Dasyueshan in Tainan, I saw a simulation of catching and releasing wild birds that presented interesting ritual aspects in that it involved actual relations of caring for birds. The Wild Bird Society of Tainan had caught three black-faced spoonbills the night before I came to visit them. These birds, visible only in East Asia, are classified as an endangered species due to the destruction of their feeding sites on the migratory pathways. In the 1990s, their number was estimated to be 2,000, but thanks to the conservation efforts among Japan, South Korea (where they breed during the summer), Taiwan, and Hong Kong (where they migrate in the winter), their number increased to 3,000 at the end of the 2000s. The Mai Po marshes in Hong Kong and the Taijiang National Park in Taiwan (north of Tainan) are two major feeding sites on the migratory pathway of black-faced spoonbills, as well as the Fujian coast in mainland China. Black-faced spoonbills were survivors from the threat of extinction.

In winter 2002–3, seventy-three black-faced spoonbills were found dead from botulism in Taijiang National Park. Botulism is a major epizootic for migratory sea birds, particularly enhanced by their concentration in wetlands where they find shelters on their flyways.[36] The Wild Bird Soci-

FIGURE 5.2. Decoys used to trap black-faced spoonbills in Tainan. Photo by Frédéric Keck, April 2013.

ety of Tainan consequently organized a campaign to vaccinate spoonbills against botulism. They used decoys to train staff on how to safely handle the birds while vaccinating them. Since then, drills are regularly organized so that bird protectors learn how to manipulate fake spoonbills with caution. These decoys are also used to attract spoonbills for GPS monitoring to follow their migratory trajectory. Carved in wood and painted in black and white, they are posted in the wetlands close to traps.

On the morning of April 29, 2013, I saw three black-faced spoonbills, that had been captured the night before, next to ten wooden decoys that had been posted in the marshes to attract them into traps. Taiwanese birdwatchers told me I was lucky; they had not been able to catch any black-faced spoonbills over the last few months, and now they got three on the same day.[37] As it was very sunny that particular morning, birdwatchers had found shelter in a Taoist temple on the banks of the marshes. They turned on some Buddhist music to soothe the birds. With very quiet gestures, a man equipped with gloves and masks sewed the satellite track around the waist of the spoonbill while a woman held it gently but firmly. Five other

FIGURE 5.3. The release of black-faced spoonbills by birdwatchers in a wetland reserve in Tainan. Photo by Frédéric Keck, April 2012.

birdwatchers watched them with amusement, taking pictures and making comments on the bird's reactions. At one point, the spoonbill managed to escape and fled to the corner of the temple, where it defecated heavily. But they re-caught it and continued sewing on the equipment. The man explained to me that he had to be very cautious because it was a young spoonbill, so the weight of the satellite track had to be imposed on its body in such a way that it would not hamper its growth or unbalance its flight.[38] After the three spoonbills were equipped in the same way and banded with colored tags, they were released in the closest pond. They walked slowly to the middle of the pond, opened their wings, and flew away.

This moment was particularly moving to observe because it mixed science and religion in a gesture of environmental protection. As I noted in chapter 4, the practice of releasing birds (*fangsheng*) is common in Taoist and Buddhist rituals and raises concerns about the health of the animals released. By inserting a simulation practice into a ritual sequence, the birdwatchers of Tainan took these ethical issues into account: they cared for the well-being of the individual spoonbills, which they had transformed into sentinels for the whole species. In the same way that volunteers in the Hong Kong exercises alternately played the role of actors and dummies, the spoonbill was represented in these simulations both as a passive inert decoy and as a resisting living body.

The language of sacrifice is often used to describe the ethical dilemmas raised by humanitarian triage or environmental tracking: patients with less urgent symptoms are sacrificed for those who must receive first aid; wild animals that die because of satellite tracking are sacrificed for the sake of the whole species they help to conserve.[39] But the simulations I have observed avoided these separations. They used decoys to create extraordinary contexts in which the question of who was the actor—the bird or the human—was suspended in a quasi-ritual order. The tensions of care in ordinary life, between patients and doctors, between animals and humans, were suspended in the time and space of the simulation because they were delegated to artifacts working as decoys. As we saw in chapter 4, sentinels are not sacrificed but open a space of identification between humans and nonhumans, which I describe here as the reverse scenario of simulations in avian reservoirs.

The release of the black-faced spoonbills resonated with other exercises that I could not observe and that, in the context of the relations between Taiwan and China, mixed military and public health concerns. Since the anthrax-letters episode after 9/11, the Taiwanese CDC had been anticipating a bioterrorist attack from the People's Republic of China and was organizing exercises simulating the use of bioweapons in a public space.[40] The SARS crisis, which lasted between March and July of 2003, seemed to confirm these fears, as 668 people were infected with a coronavirus that came from China, of whom 181 died. It was considered a failure in the capacity of Taiwanese health authorities to control infectious diseases after a series of successes against malaria, tuberculosis, and dengue. The CDC consequently organized simulations of bioterrorist attacks with a SARS virus in Taipei City Hall on September 22, 2004, with a smallpox virus on April 14, 2005, and with an avian influenza virus on December 8, 2005, in the Taipei metro. The head of the Taiwanese CDC, Steve Kuo, declared to the Taiwanese media in October 2005, "In terms of preparedness, Taiwan was in a better position than in 2003, when SARS arrived." After 2005, Steve Kuo organized these exercises with the Department of Defense on the model of the U.S. CDC. The Taiwanese Department of Defense was already organizing its own exercises in the Taiwan Straits, simulating a military attack from the People's Liberation Army.[41]

The Taiwan Straits has been a site of simulation and speculation for a long time. In the seventeenth century, Ming dynasty general Zheng Chenggong (also known as Koxinga, the founder of Chinese Taiwan) prepared to invade Taiwan, occupied by the Dutch, from the Fujian coast, where he invented for his soldiers the Mooncake game (*bo bing*), adapting a classical game of dice.[42] In this game, some dice combinations are associated with specific prizes, thus giving an indication of the good luck of soldiers. A statue of Koxinga in Xiamen still commemorates the invention of this game. The island of Kinmen, two kilometers from Xiamen, has also been a site of simulations and drills between 1949, when it was invaded by General Chiang Kai-Shek and considered the space of resistance to the People's Liberation Army, and 1992, when the Nationalist army left the island after the lifting of martial law. Kinmen then became a site of natural and cultural heritage particularly appreciated by birdwatchers.[43]

The release of a black-faced spoonbill in the same Taiwan Straits becomes meaningful in that context: instead of carrying information on a potential war between enemies, it produces signals on the changes of avian habitats that concern birds as well as humans. Through this reinvented ritual, simulations of bird diseases configure a living being into a sentinel of future threats in the Chinese Sea. Simulations in avian reservoirs thus combine different logics: security (protecting against a threat coming from the border), science (producing knowledge by counting living beings), and religion (instituting identification between humans and nonhumans through collective objects). While simulations of epidemics are often traced back to Cold War exercises of preparedness, they can thus be connected to hunters' practices with animals. The final sections of this chapter will consider this double genealogy to examine the relations between play, fiction, and rituals in simulations of avian diseases. After the ethnographies of simulations practiced by public health officials and bird watchers, these two sections will be more theoretical, discussing aspects of the exercises in a comparative perspective.

SIMULATIONS OF NUCLEAR BLAST AND APOCALYPTIC SCENARIOS

Historian Patrick Zylberman, describing exercises for pandemic preparedness in the United States, talks about a "contamination of the historical world by the fictional world."[44] He thus comments upon the effect of novels such as Richard Preston's *The Hot Zone* (1994) and *The Cobra Event* (1997) in the construction of scenarios for films or drills that convinced official leaders to invest in public health preparedness. Zylberman refers to the work of philosopher Jean-Marie Schaeffer, who defines fiction as the signaling of an intention to feign, and draws a distinction between lure and simulation. While only the person who lures is supposed to have an intention, in simulation all the actors are endowed with intentions and thus can anticipate each other's reaction. We have seen how decoys used in simulations of avian diseases are transformed into actors by scenarios of preparedness. I argued in chapter 4 that the lure becomes a decoy by capturing a set of intentions in its display. Following this hypothesis, I will now describe simulations as series of actions in which actors are endowed with intentions and ask what kind of signaling is produced in the space of simulation. How do sentinels act in simulations by signaling an intention to communicate?

The genealogy of disaster simulation indicates that simulation was first used not to anticipate emerging pathogens but to tame the power of nuclear radiation. Historian Peter Galison recounts how computer simulation emerged in the field of particle physics as a subculture among experimenters and theoreticians under the mathematical program "Monte Carlo," which was invented in the 1930s and used for deterrence in the 1960s. Simulation replaces experimentation when the data needed for the theory are too numerous and complex to be processed with traditional instruments. It compensates for the sense of loss of reality that particle physicians felt when nuclear physics entered the real world, as computer images create a virtual reality on which it is possible to act within the lab. Simulation thus plays a crucial role in the collective work of building proofs in particle physics; it creates a "trading zone" in which different professional groups—electrical engineers, physicists, airplane manufacturers, applied physicists, and nuclear weapon designers—share a common language or "pidgin."[45]

This analysis is often quoted in science studies on the use of simulation in architecture design or climate models.[46] Similarly, I argue, simulations of bird diseases create a "trading zone" in which virologists, public health planners, and birdwatchers interact around the anticipation of the extinction of a species—either humans or birds. However, despite Galison's exciting promise to catch reality within science through the images of simulation, studies following his proposal often fail to demonstrate how actors engage in these practices. Simulation is mostly considered in these cases to be a "regime of signs" without clear examples of how these signs create contexts for action. It is interesting to notice, by contrast, that Herman Kahn, often mentioned as the inventor of worst-case scenarios,[47] was very precise in his descriptions of the role of actors. This American expert in futurology advised the U.S. government on how to wage a thermonuclear war through the use of scenarios, which he defined in 1962: "A scenario results from an attempt to describe in more or less detail some hypothetical sequence of events. The scenario is particularly suited to dealing with several aspects of a problem more or less simultaneously, [helping us] get a feel for events and the branching points dependent upon critical choices."[48] In the scenarios he designed for the RAND Corporation, players were divided into opposing teams representing states at war, and the game director played secondary actors as well as "nature," whose function was to introduce chance events into the game.[49]

Anthropologist Joseph Masco has thus investigated the role of animals in "borderlands" where humans prepare for nuclear blast, such as Los Alamos, the territory of New Mexico where Project Manhattan was first implemented. Animals are used in the simulation of a nuclear winter because the effects of radiation on nonhuman bodies appear as an anticipation of its effect on human bodies. Masco considers the 1954 film *Them!*, portraying gigantic carnivorous ants that result from a nuclear explosion in New Mexico, as "the Ur-text for an ongoing fascination with mutation in American popular culture."[50] In 1948, he recalls, fish were passed on photographic plates to capture radiation after nuclear tests, because all the fish in the food chain transmitted radioactivity "like an epizootic."[51] In 1957, Operation Plumbbob, conducted in the Nevada Proving Grounds, involved 135 pigs placed in individual aluminum containers, shaved to simulate human skin, painted with various materials, and exposed to the nuclear blast. Masco notes that the use of animals as instruments for radiation research in open air ceased after the Limited Test Ban Treaty in 1963. Animals now appear as environmental sentinels for nuclear radiation in sites where their effects on humans are still contested, such as Chernobyl and the Pueblo territory adjacent to Los Alamos.[52]

After the Limited Test Ban Treaty, however, simulations of nuclear explosion had to be practiced indoors and consequently became highly secretive, producing a language of initiation that separated scientists from ordinary citizens but also humans from nonhuman animals. Hugh Gusterson has argued that nuclear simulations in these closed settings can be compared to rituals because they "alleviate anxiety and create a sense of power."[53] This gives an indication of the difference between desktop and real-ground exercises. Desktop exercises such as "Great Wall" are rituals in the sense defined by Gusterson because they lead politicians to stabilize the pandemic threat through a computer game that mimics decision-making. But real-ground exercises in hospitals or public buildings are confronted with the differences between humans and animals in their modes of contagion and engagement. In between, simulations organized by Gavin Smith in Singapore or by birdwatchers in Taiwan and Hong Kong play on the instabilities of relations between humans and birds to perform the threat of a pandemic or bird extinction.

Historian Tracy Davis thus uses the notion of performance rather than ritual to describe the techniques invented by the world of civil defense in the United States and the United Kingdom after 1945. She shows that these techniques of performance came directly from the world of theater. "Theater (and not merely spectacle)," she argues, "had a utility in twentieth-century governance, education and social life, central not only to how anxiety was expressed, but more importantly to how people envisioned ways to identify and resolve anxious problems."[54] Indeed, people coming from the world of theater could implement scenarios built by Cold War experts with a high degree of realism through the design of accessories.[55] Simulation designers were, therefore, less concerned with the participation of the public than with the realism of the artifacts composing a world in which the coming catastrophe seems to be realized—what is called in the language of theatre "accessories." Being prepared means not only imagining the nuclear blast, but also rehearsing proper behaviors with adequate artifacts in such a way that the impossible becomes real. As Davis notes, "It is not 'performance' that matters here but the preparation for it: namely, rehearsal and what was accomplished through it."[56]

Tracy Davis shows that animals played a decisive role in the simulations of civil defense because, as companions of daily life, they were an important element of realism that conveyed the sense of destruction brought by a nuclear attack. After an exercise organized in Coventry in 1954, Prime Minister Churchill complained about the high degree of realism of the simulations, which he considered a waste of public money. "Who thought of the blood-stained old woman with the birdcage? I hope there is not going to be any more of this sort of thing at Government expense."[57] The actor, notes Tracy Davis, was a member of the Casualties Union who had played the same role with the same bird during a previous exercise in London. Contestation of the realism of the simulations came not only from governments but also from participants themselves. During an exercise in New York in

1959, a young woman with her screaming child "was ordered to take cover but refused, saying 'I do not believe in this. . . . This is wrong.'"[58]

Tracy Davis brings to the front the gender division that, as we have seen with the paternalism of discourses of simulators in Singapore and Hong Kong, is a major component of the tensions within these spaces of interaction. Civil defense was defined predominantly as a woman's task because women were in charge of the household and could control its more minute details—including the behavior of pets. The role of civil defense was to bring home the knowledge of a nuclear attack in the same way as physicians had taught women the language of microbes since the beginning of the twentieth century.[59] Davis mentions a civil defense recommendation: "Women should instruct their children about what to do in the event of atomic attack, practice controlling and containing their pets, reduce fire hazards, learn first aid, and memorize the alert signals."[60]

After the end of the Cold War, these techniques of performing disaster were transferred to the world of public health management with the same asymmetry between positions of power. While the organizers of simulation are predominantly elderly males, the actors are mostly young and female; between these two levels of scenarios, pets, birdcages, and tags appear as intermediary objects of care and concern, introducing degrees of realism in the collective imagination of the coming threat. The public performance of disasters stages relations between persons and artifacts, humans and animals, elderly and young, men and women in such a way that they can be manipulated and played upon. The notion of play is therefore central to understand how virus hunters and birdwatchers simulate an apocalyptic bird disease.

PLAYING WITH PREDATORY RELATIONSHIPS IN RITUAL PERFORMANCES

Following the general argument of this book, simulations should not be analyzed only as pastoral techniques, mobilizing populations under a common threat in which some are sacrificed for the sake of others, but as hunting techniques, instantiating relations of identification between humans and animals around a perceived uncertainty at the borders between species. Gambling and playing are essential parts of the writing of worst-case scenarios, but they are oriented by the script of a performance. The multiple levels of signaling that I identified in chapter 4 now become entangled into one sequence of action. The possibility of luring or being lured, a key aspect of sentinel devices, is captured in the decoys and other accessories used in simulations. I now want to support this analysis of simulations by the anthropology of rituals in hunting societies.

As we have seen in chapter 1, Claude Lévi-Strauss developed a reflection on cannibalism across all his career and followed the works of Carleton Gajdusek in Papua-New Guinea between the 1950s and the 1990s. He gave a course at the Collège de France in the academic year 1974–75 titled "Cannibalism and Ritual Dressing Up (*travestissement*)," in which he defined cannibalism as the inferior limit of predatory relationships and communication as the superior limit. He endorsed this hypothesis with rituals practiced by cannibalistic societies, in which, he noticed, women played an ambivalent role, being either excluded from cannibalistic meals or placed at a predominant position at the banquet. Lévi-Strauss then borrowed Gregory Bateson's analysis of the Naven ceremony among the Iatmul in Papua-New Guinea. In this ceremony of initiation for young men, the classificatory mother's brother dressed up as an old woman making jokes and telling gossip, while women acted as male head hunters. This case opened up for Lévi-Strauss a compared anthropology of rituals in which the positions of men and women are reversed to simulate the conditions of war that gave birth to society. Among the Pueblo of North America, for instance, "the theft of brides, considered a hobby for young men when they were not engaged in military operations, shows that the inner life of the groups, when it is separated from the structural factors that could entertain its dynamism, was reduced to simulating the serious fights displayed on a larger scene."[61] Lévi-Strauss concluded his course by saying that ritual dressing up, rather than expressing a symbolic revolt of women against men for their domination, was a way for men to simulate the conditions of war through the mediation of women.

Lévi-Strauss later seemed to abandon his analysis of cannibalism to open a field of research on the "house" (*maison*), even though he used it to diagnose mad cow disease in the 1990s. But these ideas were brought to fruition by the next generation of researchers who insisted on the role of animals and artifacts in cannibalistic rituals. In *Spears of Twilight*, Philippe Descola has shown that the relations between groups at war in the Jivaro area are linked to the relations between predator and prey when hunting animals, which can be captured in the shrunken heads (or *tsantsa*) circulating between groups as ritual connectors.[62] In their reading of Bateson's ethnography of the Naven ceremony, completed by other ethnographies of the Iatmul society, Michael Houseman and Carlo Severi analyzed the multiplicity of relations that are displayed in this ceremony, particularly totemic relations between humans and animals (pigs, crocodiles . . .), to criticize the cybernetic model in which Bateson tries to encompass them. For Houseman and Severi, the proliferation of these relations in daily life produces contradictions, and ritual allows participants to condense and manipulate these contradictory relations by staging an extraordinary context for speech and action.[63]

In an article entitled "Dissimulation and Simulation as Forms of Religious Reflexivity," Michael Houseman, following on the theory of ritual action he has developed with Carlo Severi, proposes to move away from a definition of ritual as a collective action based on a shared secret to study "the emergent properties of ritual simulation generally." He draws on two cases taken from hunting societies of Central Africa, one in which male novices are symbolically killed and the other in which a goat is actually sacrificed. The question he asks is, how do participants come to believe in a series of speech-acts—"the novice is dead" and "the goat consents to sacrifice"—that are obviously false? His answer is that ritual action creates a space of reflexivity on the relations that constitute the group, where individuals can take on other perspectives about these relations than they ordinarily do. "What is certain for both categories of participants, beyond the painful and anxiety provoking character of the interactions they are engaged in, is the perception of each other's behaviour in relation to their own. It is this social perception, I suggest, rather than any definite conceptualisation of the episode in question (is it 'real' or 'simulated'?, what is its 'true' meaning?, etc.), that provides the grounds for their personal commitment to the genuine effectiveness of this event."[64]

While the ritual for male novices is a "dissimulation," in which the relations between initiated and uninitiated are constantly reversed in such a way that the moment of death becomes uncertain, the ritual of goat sacrifice, says Houseman, is a "simulation" of consent, in which the silencing of the goat through the consumption of herbs mimics the consent of male hunters to their initiation. The ritual is reflexive in the sense that it stages the condition of the ritual itself—the secrecy known by initiators and unknown by the initiated—through the manipulations of daily relations between humans and nonhumans, whose asymmetries are reflected as in a mirror. Houseman concludes, "What are simulations for experienced initiators remain dissimulations for less experienced ones."[65] Learning how to act through ritual action entails shifting from dissimulation to simulation, from the concealment of a secret to the inscription of one's actions in reflexive circles of relations.

The cases of simulations of avian diseases I have related in this chapter appear as extraordinary rituals only if they are thus reversed as in a mirror to display the multiplicities of ordinary relations they contain. Scenarios of bird-generated epidemics play on the asymmetries between humans and birds and the uncertainties of their encounter, encapsulating them in artifacts such as decoys, caps, tags, and dummies. If actors of these simulations are aware that the reality of a pandemic coming from birds is uncertain, they act in such a way that their relations with birds and among themselves are displayed in the present, connecting them with a set of other relations involving gender and ethnicity. In this chapter, I have made a distinction between rituals and performances, or between desktop exercises and real-ground exercises, but it must be encompassed by the distinction between dissimulation and simulation. These are degrees of reflexivity in the playing out of relationships around a secret that all share but no one is able to face: the moment of emergence of pandemic avian influenza is impossible to predict.

I propose therefore a distinction between mimetic play and reflexive ritual. If virus hunters and birdwatchers play as if they were birds—by following the movement of bird pathogens on a screen or by simulating the flight of a bird in a game—ritual encompasses these games in a reflexive framework, where all the relations between actors of the game can be manipulated. The form of "reverse scenario" displays these relations in such a way that they can be acted upon. This distinction between play and ritual has been clearly made by Roberte Hamayon in her ethnography of Siberian shamanism. In this chapter, I have analyzed simulations from East Asia, America, New Guinea, and Africa, but the practices of Siberian shamans also shed light on aspects of these simulations that have remained unexplored.

At the beginning of his influential book on possession in Ethiopia, Michel Leiris quotes this depiction of shamanism by Jean Filliozat, a specialist of India teaching at the Collège de France: "The practices of Siberian shamans are frequently accompanied with frenzied dancing to the sound of a drum, which is one of the most typical accessories for shamans. It has been much discussed whether it was a violent form of possession or a simulation of possession that occurred in the past and whose fixed characteristics would be ritually reproduced. The frenzied shaman indeed doesn't proceed randomly, everything happens as if he was playing like an actor the role of a madman in which the unfolding of gestures and speeches would be fixed in advance."[66] Leiris applies these analyses to his ethnographic observations in Ethiopia. When the Ethiopian *zar* is possessed by a horse, the same suspension of belief happens as when a Siberian shaman dances like a bear: the faked identification is an opportunity for all participants to display their relations. And yet there is something more in shamanism than possession: the perception of the outcome of future hunts drives the shaman toward spirits, rather than being passively possessed by them. Simulation actively imitates an animal so as to mitigate the uncertainty of catching it. Leiris's reading of shamanism makes an important shift in anthropological theory from simulation as a lying dissimulation to simulation as a reflexive play; but he doesn't go far enough in the analysis of relations between humans and nonhumans involved in the simulation.

Roberte Hamayon has proposed an extended reflection on the various uses of "playing" (*jouer*) in the hunting societies of Mongolia and Siberia. The Mongolian term for "playing" means "jumping like domestic animals"[67]—a meaning that has been lost in Western thinking because of the criticism of playing as a childish relation to the world. The role of the shaman, Hamayon argues, is to imitate well the animals that hunting is supposed to bring to the society. When shamans imitate animals, they follow on their movements in another scene than daily life. "The shamanic ritual also reinvents by mimicking. The shaman's gestures are also inspired by the same animal models; but the imitative mode differs: it *includes* the future developments of the animals' movements. Therefore I will call the shamanic ritual a simulation."[68] If, as Rane Willerslev argues, taking shamans seriously implies entering the life-world in which they are and are not animals at the same time,[69] then, Hamayon argues, it is necessary to understand how fiction enters the life-world of shamanistic societies. These societies need fiction because they have to cope with the uncertainties of the hunt and anticipate their future encounters with animals. Simulation is a dimension of fiction necessary for hunters to act upon a constantly mutating reality. While hunters are mostly men, shamans are mostly women because they reverse the relations between humans and nonhumans through the performance they play with their drums, in which these relations are reflexively displayed as in a mirror.

A word in English captures this uncertainty met by hunters and performed by shamans: "game," in the double meaning of what the hunt aims at (killing wild animals) and its mode of action (playing like an animal before killing it). Another word captures this double meaning: "chance." A hunter invokes chance if he/she wants to have a good catch, but he/she cannot calculate its probability (in contrast with hazard). Game and chance,

Hamayon notices, are partitive notions: it is impossible to separate parts in this vague ensemble, but it is necessary to redistribute them to sustain a favorable future. This explains why lucky hunters have to redistribute their hunt at the risk of being punished by the spirit of the animals. No statistical knowledge based on probabilities is necessary to produce the mutual reasoning that allows hunters to distribute the game. This taming of the uncertainties of the future through collective imagination is what Hamayon calls being prepared. She writes, "In order to grasp the specificity of play, we must first of all ask ourselves if all that is 'to come' has to be 'prepared.'"[70]

When they simulate the transmission of diseases from birds to humans, virus hunters and birdwatchers display the problems arising from the distribution of values produced by their relations: pathogens, vaccines, antivirals, but also bird specimens and images emerge as artifacts in this chain of relations. Simulation is analogous to ritual in that it dissimulates the fact that preparedness produces inequalities between the actors of the management of avian influenza. The realism of worst-case scenarios is questioned because simulators are aware of the asymmetries between the actors who participate in these performances—what Leiris called after Sartre "bad faith" (*mauvaise foi*). And yet they can only multiply forms of reverse scenarios to approach an escaping reality through new simulations. Realism is a moral injunction that impulses the game of fiction writing, not a stable position in an epistemological battleground.

We have met the questions of trust and criticism in the chapter on sentinels and the questions of reality and secrecy in the chapter on simulation; we now have to raise the questions of justice and equity in the management of bird flu. If sentinels produce communication between humans and animals at the risk of being lured, if simulations engage actors in virtual relations between humans and animals at the risk of falling into fiction, we can now see how techniques of storage and stockpiling prepare for an uncertain future by stabilizing inequalities in the present.

CHAPTER SIX

stockpiling and storage

This chapter looks at the accumulation and exchange of bird flu samples as modes of regulation of the global circulation of influenza viruses. Stockpiling, as a technique of preparedness for future catastrophes, is also a crucial element of the economy of emerging infectious diseases. If it is impossible to predict where and when a pandemic virus emerges, it is possible to stockpile treatments (vaccines and drugs) that will mitigate the threat when it appears. The imagination of a future catastrophic event leads public health managers to reorganize what is already there in the light of what is about to arrive. Stockpiling, I will argue, produces a new form of biovalue in the sense of a relation between living beings and a future state of the world.[1]

To make this argument, I will contrast stockpiling with two correlated notions: storage and livestock. While stockpiling produces value in relation to an anticipated future, storage produces value in relation to a conserved past. Viruses are stored to know about their past mutations and attenuated to make vaccines, which are stockpiled for the next pandemic. Livestock, by contrast, refers to an uncontrolled reservoir of viral mutations produced by the accumulation of live animals. It appears as the dark face or the "accursed share" of stockpiling.[2] While stockpiling is oriented toward the mitigation of an emerging pandemic flu virus, livestock is retrospectively perceived as the origin of a new virus. Storage, by contrast, seems to be practiced for its own purpose, with the idea that any virus stored can help to understand the emergence of a pandemic. Storage, stockpiling, and livestock appear as three ways to accumulate and exchange viral forms in relation to a pandemic event.

To discuss these relations between structure and event, I will refer to the anthropology of hunter-gatherers. Similarities between the practices of virus hunters and birdwatchers will be extended to their practices of accumulation and exchange, particularly in the use of databanks that produce knowledge on the vulnerabilities shared by birds and humans. The distinction between hunting techniques of preparedness and pastoral techniques of prevention will thus be put to the test of the economy of relations. If storage is a practice of hunting societies following animals through their virtual traces, how is it reconfigured as stockpiling by its orientation to a pandemic event? Ontologies of relations between humans and animals and techniques of anticipation of natural disasters will thus be linked to modes of production of viral information.

STOCKPILING VACCINES, STORING VIRUSES

On April 23, 2013, I visited the Animal Health Research Institute of Taiwan, situated on the banks of the river Danshui in the north of Taipei. Danshui (meaning "fresh water") was a familiar landscape for Chinese and Japanese seafarers approaching Taiwan and a site of settlement for the first Spanish and Dutch explorers in the seventeenth century, who built and occupied a military post overseeing the straits between Taiwan and China, today known as Fort Santo Domingo. The British established their first consulate here in 1860 when the treaty of Beijing opened Taiwan to Western trade, and Robert Swinhoe arrived in 1862 to administer the Chinese Custom. It thrived with the export of oolong and baozhong tea by Jardine Matheson until the beginning of the twentieth century, when it was replaced by Keelung as the main harbor for north Taiwan. The Animal Health Research Institute was part of an academic campus including the University of Aletheia ("truth" in Greek) and Oxford College, created by George Mackay, a Canadian Presbyterian missionary born in Oxford, Ontario, in 1844, who arrived in Danshui in 1872, married a local woman, and developed medical education until he died in 1901. I was at the heart of one of the most connected colonial posts of the island.

The first human case of avian influenza in Taiwan had just been declared. A businessman returning from Shanghai was identified as carrying the H7N9 strain that had infected more than a hundred people in China in the preceding months, killing one out of five. Since the 1997 emergence of the H5N1 virus in Hong Kong, there had not been any human case of bird flu in Taiwan. However, in 2013, the H7N9 virus, although less virulent, seemed to be spreading more rapidly in China. Despite intense pressures to contain the virus and communication constraints on public health officials, the head of the Animal Health Research Institute, Hsiang-Jung Tsai, agreed to meet me and explain practices of surveillance of flu viruses in animals.[3]

The H7N9 virus is not new, he told me. It had previously been found twice in wild birds in Taiwan, although each time with a different genetic sequence. For the past ten years, the Chinese Wild Bird Society of Taiwan had been collaborating with the Animal Health Research Institute to define areas of study from which they collected bird feces. At the time of my visit, 50,000 samples had already been collected in 2013, out of which 3,000 were found to contain various influenza strains. A monitoring program was also conducted on domestic poultry in farms and markets in Taiwan. In May 2012, around 4,000 cases of H5N2 (not transmissible to humans) in poultry farms were declared to the World Organization for Animal Health (OIE).[4] But rumors said that the same strain had caused many more cases between December 2012 and February 2013, which had not been declared because of the campaign for the presidential election. This rumor was confirmed by a study of the National Taiwan University showing that the H5N2 strain had been introduced to Taiwan in 2003 by a vaccine produced in Mexico.[5] In July 2012, when smuggled wild birds with H5N1 were found at Taoyuan Airport on a plane coming from Macau, they were immediately destroyed.[6] Even if avian influenza had never killed any person from Taiwan, the surveillance of its mutations in animals was a sensitive political question, which revealed the tense relations between Taiwan and mainland China.

In November 13, 2013, it was reported that a twenty-year-old woman had been infected the previous May by a new H6N1 bird flu virus.[7] The news was announced only in November because researchers from Taiwan Centre for Disease Control had first to sequence the flu virus from the woman (who had recovered in the meantime), then search and sequence analogous flu viruses in samples collected from chickens. The Taiwan Center for Disease Control ultimately showed that seven of the genes found in the woman were closely related to a flu strain isolated from Taiwanese chickens that year. The eighth gene was most closely related to another strain first found in Taiwanese chickens in 2002. It was uncertain how this woman came into contact with chickens, since she was a clerk at a deli, but a "molecular clock" analysis, which tracks the evolution of mutations in a genome, made it almost certain that the flu she had contracted had previously circulated among birds. In 2013, this knowledge aroused fears that it could jump again to humans, with more catastrophic effects than in 2002.

How could you know, I asked Hsiang-Jung Tsai, that it was the same H7N9 that circulated in humans and birds? As an answer, he showed me a fridge, with the temperature -80°C glowing in a red digital thermometer.[8] He told me that 200 wild bird flu viruses were stored in that fridge, two of which were H7N9. For the first, collected in 2009, they sequenced the HA genetic segment, which indicates the virulence, and for the second, collected in 2011, they sequenced eight other segments. Through the efforts of sequencing and comparing mutations in each strain, they could thus show that the 2013 human strain derived from these wild bird strains. The conservation of viral samples in a fridge made possible their presentation through a phylogenetic tree. As Hannah Landecker writes, freezing techniques, when they were introduced to cultivate cells in the 1950s, synchronized a world of living beings characterized by mutations and turned them into objects of manipulation—or, in Landecker's term, cultivation. "The freezer acted as a central mechanism both within individual laboratories and companies and within the biological research community more generally to standardize and stabilize living research objects that were by their nature in constant flux."[9] Landecker also notes that this technique was applied from the stabilization of human cell lines in the lab to the control of sperm cells for cattle breeding.

In another part of the Animal Health Research Institute, I was shown fridges containing vaccines for poultry. These were conserved at a temperature of +4°C, which is higher than virus samples, because, I was told, it was sufficiently cold to preserve adjuvants. Adjuvants are proteins that boost the reaction of immune cells. Following a technique implemented in the 1960s, flu vaccines are live flu viruses cultivated in chicken embryos and attenuated so that they present their antigens to the immune system, which, in return, produces antibodies. The production of flu vaccines is an industrial challenge, since the flu strain circulating in a population mutates constantly. Every year, influenza experts gather in symposia called by world health administrations (WHO in Geneva for humans, OIE in Paris for animals) to collectively agree upon which strain should be the target of vaccination efforts. But for humans, the ordinary production of vaccines against seasonal flu sometimes interferes with the extraordinary production of vaccines against pandemic flu coming from animals.

Edwin Kilbourne, a leading virologist at the Rockefeller Institute in New York, was the head of the U.S. Strategic National Stockpile until his death in 2011. This repository of antibiotics, vaccines, and other medical equipment was created in 1999 by the federal CDC under the name "National Pharmaceutical Stockpile." In 2004, Kilbourne advocated the production of a vaccine reassorting the thirteen known influenza A virus subtypes to act as a "barricade vaccine" before a "rampart vaccine" could be made for the actually emerging pandemic virus.[10] The main question for him, as a first observer of the "Swine Flu Fiasco" in 1976,[11] was how to stockpile vaccines in such a way that their distribution in case of a pandemic would be safe and fair. The sites where human vaccines are stockpiled are kept secret, because they could be main targets for terrorists preparing a pandemic flu attack—or looters wishing to benefit from the panic when the pandemic occurs.

Vaccination of animals raises other economic and moral issues, which explains that the stockpiles of animal vaccines are much more accessible. According to the recommendations of the OIE, vaccination of domestic poultry is a reasonable complement to the slaughtering of infected chickens, but should also be accompanied by the surveillance of the mutations of flu viruses among wild and domestic birds. While China and Vietnam have been criticized for vaccinating their domestic poultry massively with vaccines produced locally, which could have led to a selection of resistant strains,[12] Taiwan chose to stockpile vaccines and distribute them only in case of an outbreak. "We are not supposed to use vaccine, because it interferes with the monitoring by marking the poultry positive,"[13] said Hsiang-Jung Tsai. "But if there is an outbreak and the stamping out policy is not enough to stop contagion, we need to use vaccine." As we have seen in chapter 1, while stamping out produces ontological ruptures between humans and animals, vaccinating and monitoring institute moral solidarity between them. But vaccination has a higher economic cost than monitoring.

The problem raised for both human and animal vaccines is: what to do with the vaccines produced in excess for strains that are not circulating anymore? In 2013, the Animal Health Research Institute of Taiwan maintained 10 million doses for H5 and 5 million for H7, the two most currently circulating strains in poultry. They were bought from French (Meyrieux), Italian (Fluvac), and Mexican (Avimex) pharmaceutical companies. After eighteen months, any non-used vaccines were incinerated, and updated vaccines were bought. Members of the Taiwanese Parliament complained about the quantity of vaccines destroyed, which led the Taiwanese government to decrease the number of vaccines stockpiled. To reduce the cost of stockpiling and destroying vaccines, the government also passed a contract with private companies capable of producing 3 million doses of vaccine within a week in case of an outbreak. A Taiwanese private pharmaceutical company, Adimmune, announced that it could produce between 5 and 10 million doses of vaccines for H7N9 in six to eight weeks, and it was consequently awarded the right to develop the vaccine. It was the first time a Taiwanese company was in charge of producing human vaccines from the beginning to the end of the development process instead of repackaging other companies' products.[14]

The development of flu vaccines is thus the object of complex negotiations between public and private interests. While the Animal Health Research Institute could produce and stockpile its own vaccines, it was compelled to rely on the initiative of private companies who invent new technologies of massive and rapid production because of concerns about waste. Though some saw the consumption of vaccines as a product of the whims of nature—whether or not an epidemic emerged—others interpreted it as a result of pressure from industry. "We are not supposed to produce too many vaccines, otherwise private companies will not be happy," said Hsiang-Jung Tsai. If the capacity of a Taiwanese private company raised pride about the island's sovereignty and efficacy, it also recalled bad memories. The first director of public health under the Guomingdang government was the head of a pharmaceutical company based in Shanghai; in 1946 he released the supply of 45 million antimalarial tablets from Japan and the quarantine measures implemented under colonial government to develop the sale of his own company's quinine.[15]

Restrictions on public research on vaccines in Taiwan have also been explained as the result of geopolitical constraints. These vaccines aimed to protect the Taiwanese population and could therefore not be exported to any other country. "If we could sell vaccines to mainland China," said Hsiang-Jung Tsai, "public production of vaccines would be profitable, but for now that's not possible. And we are not allowed to buy vaccines from mainland China." It must be recalled that Taiwan had difficulties joining the World Health Organizations (WHO and OIE) due to the opposition of mainland China and has been admitted only in 2009 under the name "Chinese Taipei." But Taiwan is also an important actor in the global pharmaceutical market: it can sell vaccines and drugs to the twenty-six countries who recognize the "Republic of China." The circulation of animal vaccines, or lack thereof, is a material expression of complex geopolitical tensions. Taiwan started stockpiling vaccines against smallpox in the 1970s, after the Beijing government had refused to sign the international convention for nonproliferation of biological weapons in 1972, at a time when a smallpox attack was dreaded on the Taiwan straits.[16]

Taiwan has also been one of the first countries to stockpile antivirals for influenza.[17] Used after infection, these drugs inhibit the neuraminidase that controls the entrance of the virus to cells. They include Oseltamivir (used to produce Tamiflu, orally prescribed) and Zanamivir (used to produce Relenza, inhaled). While vaccines need to be constantly updated and recycled, antivirals can be kept for five to seven years at a low temperature (2 to 8°C). In Taiwan, the Department of Health claimed it had stocked 2 million doses of Tamiflu, providing treatment for 10 percent of the population. These doses were mostly produced by Roche, a Swiss company that launched the antiviral ten years ago and then became the global provider after 2005. In July 2005, the Taiwanese government donated 600,000 doses of Tamiflu to the government of Vietnam, which was facing a major outbreak of H5N1, to stop the spread of the virus on the whole Asian continent.[18] In November 2005, the Taiwanese Department of Health questioned the company's capacity to produce enough antivirals for the Taiwan population in case of a pandemic and threatened to develop the antivirals with its own license; but Roche finally issued the 2 million doses required in 2006.[19]

The situation of Taiwanese health authorities in regard to stockpiling viruses, vaccines, and antivirals can be contrasted with that of Hong Kong. The Special Administrative Region is probably one of the biggest consumers of Tamiflu, due to its long exposure to emerging flu viruses and its mobilization on the threat of pandemic flu. Wikileaks reports that 20 million doses of Tamiflu have been stored in Hong Kong, which is the highest rate of coverage in the world (300 percent), while Western countries cover between 20 and 50 percent of their population.[20] This may explain why flu strains resistant to Tamiflu appeared on the territory in 2009 after the antiviral was massively used.[21] Roche is regularly mentioned in the Hong Kong newspapers when there is a risk of shortage of Tamiflu, generally at the entrance of the cold season.[22]

When I took part in a virology congress in Hong Kong in March 2009,

I attended a private workshop organized by Glaxo Smith Kline, the competitor of Roche for antivirals producing Relenza, to examine which flu strain was the most serious candidate for pandemicity. The organizers claimed that whatever new strain would emerge, there were enough antivirals and vaccines to mitigate the pandemic threat. The only debate was to examine the different "candidate viruses" for the next flu pandemic. We have seen in chapter 1 that the vaccines used in poultry farms came from the Netherlands and Mexico, but that the growing inefficacy of these vaccines raised discussions on the opportunity to buy vaccines from mainland China, where new strains appeared and where pharmaceutical companies proposed good prices. During the H1N1 pandemic in 2009, the Hong Kong Health Department bought only French vaccines, after a long deliberation during which they were compared to their Chinese counterparts. As a radically liberal economy governed by the fear of a pandemic coming from mainland China, Hong Kong was a good territory for pharmaceutical companies to compete.

The global circulation of viral strains and vaccines raises issues of sovereignty that were clearly staged when the Indonesian Public Health Department refused to share with WHO the H5N1 strains that circulated in its population if it was not granted preferential access to vaccines produced out of them.[23] While Taiwan tries to preserve its viral sovereignty in front of bird flu strains coming from mainland China, Hong Kong increasingly assumes its dependency on mainland China not only for imported poultry (with the flu strains that accompany them) but also for vaccines. It compensates this dependency with another dependency on the U.S. company Roche and the European company Glaxo Smith Kline for the production of antivirals. The Indonesian case showed that the production of treatments for flu pandemics is an unequal gift-giving relationship. "Southern" countries give strains and receive treatment from "Northern" countries having the technical capacities to produce vaccines and antivirals. But the relationship between Taiwan, Hong Kong, and China is more complex than this North/South opposition. If China has both the strains and the capacity to produce treatments, Taiwan and Hong Kong have to invent new forms of sovereignty in the margins of China, because the circulation of commodities, persons, and virus strains is particularly intense in these margins. There are thus many forms of gift-giving relationships in the world of flu viruses and many places where these relations become more reflexive and speculative.

Hong Kong microbiologists have found an interesting solution to this impossible equation. Instead of producing vaccines and antivirals, which they would give to the rest of the world, they produce knowledge on the emerging strains to raise alarm for the rest of the world. A sentinel post is not only a liminal place on the global battlefield in the fight against emerging infectious diseases; it is also a profitable position in a global economy of knowledge. Remember the work done by Gavin Smith and his team in Hong Kong and then in Singapore, dating the emergence of pandemic flu strains through computational simulations. When he mentioned the fact that the access to Shantou where they could collect bird flu samples had been cut off by the Chinese government, Vijaykrishna told me:

> Closing Shantou is basically a shame. All the information that we've had in the past ten years is because of this massive surveillance that's been going on in this region. And when the H5N1 virus spreads, we could say: "this is the genotype that is spreading." WHO could be preparing, we could send vaccines to these countries where the virus is endemic. But since 2006 when Shantou was closed, we don't know much about this region, so we suppose these nucleotides are spreading outside. Hong Kong is the perfect example of how transparent it can be. Surveillance in poultry, swine, wild birds, as soon as it's done, it's made public. And that doesn't mean Hong Kong is a bad place, it means that everything that is done is reported. We're sitting on a bunch of information. Viruses are there, still unknown. Even if we don't do surveillance, we have enough information to work for five years.[24]

Vijay meant that even if the access to actual flu strains were cut, his team of virologists could still work on a virtual biobank, considered a repository where data waited to be valorized through knowledge.[25] The large number of strains collected by Shortridge, Webster, and all the virus hunters they had trained were now stored in the University of Hong Kong or at St. Jude Hospital in Memphis. They were then sent to the website of GenBank at the U.S. NIH, where virologists could retrace their history using computer software.

Vijay made an important distinction between the wetlab, where viruses are identified, purified, and frozen, and the drylab, where they are sequenced, compared, and aligned. "In general, I don't do the lab work, everything that makes your hands dirty. I work mainly on computers. But if I didn't have the staff to do the sequencing, I could do it because that's what I've done for years."[26] The wetlab and the drylab are two stages in the production of viral information, transforming bird feces into valuable knowledge. Viruses are dessicated and purified before being sequenced and stored. These sequences constitute a bounty of information independent from the access to live reservoirs of viruses.

Parallel to the massive amount of vaccines and antivirals stockpiled by health authorities, then, there is a massive amount of viral information stored in microbiology labs. While stockpiling is ruled by sovereignty, property, and exchange, storage is regulated by trust, transparency, and robustness of the information. How to describe this difference between stockpiling and storage? It is interesting to notice that there is only a difference of temperature between stockpiling antivirals (8°C), stockpiling vaccines (4°C), and storing viruses (-80°C). But we need to explore this difference both epistemologically (what kind of knowledge do they produce?) and economically (what kind of value do they produce?). The difference in temperature between stockpiling and storage may amount to historical ruptures between modes of production.

COLLECTING BIRD SPECIMENS AND IMAGES

To analyze the storing practices of microbiologists, I want to turn once again to those of birdwatchers. I will trace debates about storing specimens in museums and storing bird images in databanks to ask about the meanings of the distinction between storage and stockpiling.

As I have recalled in chapter 3, ornithology is a Western practice, dating from the eighteenth century, based on the comparison of birds collected all over the world in a central museum, and compared to birds observed in the wild on a local patch. We have seen in chapter 4 that this practice was appropriated by the Chinese middle class at the end of the twentieth century to promote environmental awareness in a context of massive urban development. How has this transfer of birdwatching from Europe and the United States to China transformed storing practices? We must once again turn to museums before analyzing more recent developments in digital databanks.

Historian Fa-ti Fan recalls that in the middle of the nineteenth century, "Robert Swinhoe employed 'a vast number of native hunters and stuffers' when he was stationed in Taiwan. The naturalists also encouraged the local Chinese—farmers, fishermen, or whatever their occupations might be—to bring them any animals that were interesting or unusual."[27] At the same time, Father Armand David was traveling in Central China to collect and name specimens of mammals, birds, and plants for the Museum National d'Histoire Naturelle in Paris. In his notebooks, he complains about the cruelty of Chinese peasants and hunters who destroy the habitats of these species.[28] In 1905, Reverend Hardy noted, after visiting the museum of zoological specimens in the City Hall of Hong Kong, "So many beautiful birds labeled to belong to Hong Kong are in the museum and so few outside that it would almost seem as if they had nearly all obtained the immortality of stuffing."[29]

The introduction of birdwatching into China can thus be seen not only as a Western concern for nature preservation but also as a museological form of conservation projected on a new space. While Chinese scientists had developed several ways to represent nature and act upon it, among which they included Jesuit techniques of astronomy, Western observers introduced techniques for mapping nature that created new needs and desires. Fa-ti Fan argues that "British naturalists saw China as a space to explore and to map. The Chinese, their will, and their society and political institutions, in this view, represented obstacles to the naturalist's quest of complete knowledge of the flora, fauna and geology of the land."[30] Mapping the Chinese territory was a way to take power over its resources and population; but this form of representation and appropriation of nature could be shared with local elites. When members of the Factory of Canton thus proposed to create a British Museum in China by at the end of the 1820s, they declared, "Considering the taste of the Chinese for live Birds and Fishes, we may hope that the richer Classes may acquire a taste for the same Animals when prepared for the Cabinet."[31] Indeed, Wang Shuheng, who had worked with Father Armand David to collect specimens in Central China, became the taxidermist at the Shanghai Museum founded in 1874.[32] But while Chinese hunters may have appeared as allies for Western bird collectors, they were also seen as obstacles by institutions organizing zoological expeditions in China, such as the American Museum of Natural History in 1916–17.[33]

When Japan included Taiwan as a part of its empire under the terms of the Treaty of Shimonoseki in 1895, a specific form of Japanese ornithology developed based on the wealth of species endemic in the tropical island. Japanese scientists didn't separate hunting from collecting and showed respect for the knowledge and crafts of Aboriginal people who knew where to find birds and how to compose motifs with their feathers—particularly the feathered costumes and headdresses of Ami people, the largest Aboriginal group living in the east of the island.[34] Japanese collectors brought many Taiwanese species to the institute of ornithology, created by Prince Yoshimaro Yamashina in Tokyo in 1932 as a private collection with 16,000 specimens. In 1984, this institute has moved to Abiko, in the prefecture of Chiba, with its 59,000 specimens and 18,000 books. Since then, it has developed a massive databank based on the DNA analysis of bird species from all over the world. The *Journal of the Yamashina Institute for Ornithology*, created in 1952, is still a main reference in global ornithology.[35]

The first field guide on birds of Taiwan was the result of a collaboration between a Taiwanese engineer trained to taxidermy in Austria, John Wu Sen-Hsiong, and a Japanese painter. John Wu Sen-Hsiong is famous in Taiwan for being a former bird hunter, a passionate bird collector, and the leader of the Taiwanese movement for the protection of the grey-faced buzzard (*Butastur indicus*). These birds breed in Japan and go down to the Philippines in the winter, passing by Taiwan in October and March. Japanese birdwatchers published reports about the decline of grey-faced buzzards, which they explained by massive hunting in Taiwan for commercial purposes. The skins of these raptors, endowed with particular virtues, were sold to Japan to serve as specimens. Articles were published in the Japanese media about the cruelty of raptor-hunting, showing nails in heads, opened skins, eyes stacked in pyramids—often borrowing from the colonial representations of Taiwanese Aboriginals as head-hunters. John Wu Sen-Hsiong, a member of the Taichung Bird Club trained in the U.S. and Japan, was met by Japanese birdwatchers in 1977. One of them was Noritaka Ichida, who worked for Wild Bird Society of Japan and BirdLife Asia, launched a series of bird books for Asian countries, and was instrumental in the success of the black-faced spoonbill international collaboration. When I met him in his workshop in Taichung, John Wu Sen-Hsiong recalled this meeting as the shifting moment of his career: "I was invited to a dinner at the Peninsula Hotel in Tokyo. There were nine persons, all famous in birding in Japan. 'We're concerned that your country ships 60,000 skins of birds of prey every year!' I said 'Your country is developed, our country is developing; we have to learn from you!'"[36]

John Wu Sen-Hsiong's mobilization from below for the protection of the grey-faced buzzards met initiatives from the top of the government, which explains its status as a success story of environmental movements in Taiwan. President Chiang Kai-Shek issued a three-year hunting ban on grey-faced buzzards in 1972 and declared it "National Day Bird," since the peak of the migration coincides with the National Day on October 10. In 1981, President Chiang Ching-Kuo created Kenting National Park, Taiwan's first national park, to protect the area of migration of grey-faced buzzards; it hosted the East Asian Bird Protection Conference in 1983. The police collected 4,000 birds trapped by hunters and released them in Kenting National Park. While the body parts of buzzards were stored as

FIGURE 6.1. John Wu Sen-Hsiong in his taxidermy workshop in Taichung. Photo by Frédéric Keck, April 2012.

elements for Chinese Traditional Medicine, the Taiwanese government, advised by Japanese experts, created natural reserves where buzzards were conserved as a national resource to prepare for future species extinctions.

Despite his nationalistic enrollment in the protection of this endangered raptor, John Wu Sen-Hsiong was most grateful to Japanese ornithologists for teaching him how to shift from a taxidermic approach to birds as specimens to a more respectful vision of birds in their environment. In his book, birds are drawn in a static fashion, following the display of taxidermic specimens mixed with the style of Japanese painting, but their precise location on maps gives an insider's knowledge on their mode of life. This small book, which appears as the printed version of a taxidermist cabinet, allows birdwatchers to go into the field and identify birds by sight. John Wu Sen-Hsiong was proud to give me his book, designed to "capture" birds in the wild, and to show me the specimens he kept in his workshop as two types of trophies from his bird hunts. He told me that the Wild Bird Society sold up to 80,000 exemplars of the book, which allowed it to support conservation projects in the West Indies and Cambodia.

By contrast, I received another gift from Lucia Liu Severinghaus, who has a PhD from Cornell University and a professorship in ornithology at the Academia Sinica in Taipei: the complete edition of three volumes of *The Avifauna of Taiwan* on a DVD. The paper edition of this encyclopedia, edited by Taiwan's Council of Agriculture's Forestry Bureau, is the result of six years of collective work and weighs eight kilograms. Its pictures follow the model set up by Roger Peterson in 1934 for the United States: birds are shown flying to facilitate identification, and specimens of the same species are displayed on the same page. Lucia Liu explained to me:

> In this book, we put more information than what is required for identification: the meaning of the scientific name, where the voucher specimen is, the history of collection, who wrote about it, measurement, general description of morphology, habitat breeding, song, and also a section on conservation, and distribution maps. So it's really a general reference book, it's not meant for people to take into the field. You can go into the field, come back and read about the species you want. But if you have an iPad, you can take the e-version into the field.[37]

The role of field guides in environmental movements has been discussed in science studies: do they raise an environmental awareness by training attention to the diversity of birds, or do they narrow the perception of the environment to images that can be presented in a book? In short, do birdwatchers read bird books to know about nature, or do they look at nature to confirm their books?[38] I would like to frame the question differently: showing how bird books are anchored in colonial practices of collecting for museums, and asking how they are transformed by digital practices today, I describe museum specimens, book images, and databanks as three forms of storage of the past to think about how they can prepare for future disasters through stockpiling.

Taiwan is a good space to raise such questions on storage because, in addition to being one of the vanguard countries in digital computing, it has recently reappraised its colonial past in the context of democratization and indigenization (*bentihua*). The role of fifty years of Japanese colonization in the creation of cultural museums and natural reserves was stressed by the Democratic Party (DPP) when it was in power in the years 2000–2008, to distance itself from the politics of industrial development of the Nationalist Party (Kuomintang or KMT). This period also saw a boom in the creation of new museums dedicated to different aspects of Taiwan as a multicultural society. In her ethnographic fieldwork on a Hakka cultural hall in Longtan (Taoyuan County), Anne-Christine Trémon quotes the head of the Yuan committee for Hakka Culture inaugurating the hall: "I expected to visit a mosquito hall (*wenziguan*) but I'm enchanted it is so good. Do not leave it empty, but strengthen the material and immaterial aspects (*ruan-ying ti*)."[39] She shows that "mosquito hall" is a term commonly used to criticize the inauguration of museums remaining empty in China, and that *ruan/ying* is a couple of notions that can be understood on the model of hardware/software as two sides of the same cultural process, one oriented toward building monuments, often sponsored by the state, the other toward providing digital information and more open to civil society. Following her argument, I want to show that museums for bird specimens and databanks for birdwatching are two sides of the same process of storing bird images to ask how they can be oriented toward the future.

I now turn to the practices of birdwatchers in Hong Kong. We have seen in chapter 2 that Hong Kong birdwatchers succeeded in distancing themselves from the colonial model of a military association to produce their own modes of knowledge on biodiversity. This may explain why the collections of natural history conserved in the Hong Kong Museum of History are restricted, by contrast with Singapore, where the collection of natural specimens is huge but birdwatching practices are restricted (if they are compared to the situation in Great Britain, where specimen collections and birdwatching practices are highly developed).[40] Mike Kilburn, former vice president of the HKBWS, told me, "We've been collecting records of birds for 50 years in HK. This gives us an authority that nobody can question on birds."[41] In his words, "records" were clearly a type of data that birdwatchers legitimately produce as an adequate image of natural diversity, by contrast with specimens extracted to nature. Mike Kilburn criticized the practices of amateur birdwatchers who take pictures of birds with fancy cameras and post them on the HKBWS website without mentioning their species or even their site of observation. He referred me to another website, Hong Kong Wildlife, which posted pictures of animals and plants without any requirement for providing information associated with these pictures. Mike argued that a picture of a bird without information on the species and the location of the finding has no value by comparison with a "record." "Birds are the greatest challenge to birdwatchers because they are mobile," he said. "Before, birdwatchers used to organize exhibitions; now they post pictures on the Internet, and 25 persons tell them 'Wow, it's a great picture.'

The HKBWS asks them to indicate the place where they took the picture, but they want to keep the scoop. There are rival groups of photographers who share the information between them but not with others."[42]

Ruy Barretto, a friend of Mike Kilburn at the HKBWS and a lawyer in a big Hong Kong cabinet, had another view of the use of images on birdwatchers' websites. For him, the value of a bird picture lay not only in the information it carries, but in the engagement it reveals for the environment from the person who takes the picture. "Birdwatchers can contribute to conservation because they are mobile, well-off, articulate," he told me. "They can make proposals and protest. If there is a conservation damage, they can photograph that. Amateurism is not bad, there is immaturity and enthusiasm, but it can provide the records from which you get the statistics and the numbers."[43] Ruy Barretto took the example of the area of Long Valley, an agricultural land on the border between Hong Kong and Shenzhen where the HKBWS has succeeded so far in stopping construction projects. On the society's website, birdwatchers are asked to visit this area where only local birds can be seen—by contrast with Mai Po, where migratory seabirds abound—and take pictures that they can share with other members. Sharing a picture, in that case, doesn't mean spreading public knowledge on nature, but participating in a collective engagement in the defense of the environment.

This controversy among members of the HKBWS on the value of photographs shows that when a new technology is introduced in the apparently disinterested and distanced practices of birdwatchers, debates about interest and appropriation reintroduce the concerns of hunters in relation to the animals they hunt. The history of birdwatchers is usually described as shifting away from the accumulation of bird specimens by hunters.[44] Birdwatchers content themselves with observing birds without catching them, which makes birdwatching an "alternative" practice in a globalized economy of commodity consumption. Yet it may be that birdwatchers introduce, in their knowledge practices, techniques that are derived from the globalized economy they seem to challenge. They transform the singular encounter with birds into data that are stored, preserved, and exchanged, raising controversies on their ownership and proper uses.[45] In the same way as microbiologists constructed biobanks allowing them to build a coherent picture of the world and make a claim to knowledge, birdwatchers have built databanks that support their claims on the protection of the environment, both locally and globally, raising the same ethical issues: who owns the data about biodiversity, the amateurs who have produced them

FIGURE 6.2. A bird amateur in Guangzhou bird market. Photo by Frédéric Keck, June 2009.

in the field or the experts who translate them through encoded norms of knowledge?[46] The debates in "citizen science" between experts and amateurs thus turn around different modes of appropriation of bird images, between storing them as reflections of natural diversity and stockpiling them as resources that can be valorized under threats of species extinction.

The former head of the HKBWS, Lam Chiu Ying, thus made a parallel between two trends in the consumption of birds: "The English say: 'Can we take pictures? Which species is it?' The Chinese say: 'Can we eat it? How does it taste?' I'm interested to know what happens when I look at a bird and it looks at me."[47] This quote allows me to formulate a hypothesis that will drive the thinking along the rest of this chapter. When birdwatchers look at a bird, they wish a reciprocal exchange of gaze, allowing them to see the world through the perspective of birds. But the birdwatching situation is asymmetrical, because birdwatchers are equipped with tools—binoculars and cameras have replaced the arrows and rifles of former hunters—while the bird has only the possibility to fly away and escape the interaction. Birdwatchers thus need to think that birds consent to give an image of themselves in return for the protection they receive in natural reserves. But when birds send pathogens to humans, it means that the frames of the interaction are deregulated and that the interaction is contaminated by potential violence.

Lam Chiu Ying expressed these tensions in his narrative of how he became a birdwatcher: he was looking at dead objects—stars—before coming to birds as living objects, but the context of his first encounter with birds—a cemetery—reminded him that birds observed as objects can turn into dead specimens. "Cemeteries are good places for birds and ghosts. It was my first birdwatching, I was 27. Suddenly it was opening a door: I was seeing the living objects. It was opening my heart. Before I was studying physics and mathematics—dead things. I was blind before. Our organs are opened but the signals are filtered out. I became an addict of birdwatching, I wanted other people to see, it was like preaching." There was something religious or mystical in birdwatching for Lam Chiu Ying, because observing birds was for him a way to exorcize the ghosts reminding humans of the violence they impose on natural beings. In the eyes of the head of the HKBWS, the economy of bird pictures must be regulated as a reciprocal exchange between the material traces of birds posted on the website and the protection they receive in their local environment. If this economy is deregulated, living birds turn into dead specimens. The singular encounter between a human and a bird is thus expanded into a circulation of information at the risk of verging on ghost hunting. In that light, birdwatching is not so far from poultry breeding, and the tensions I have observed in the economy of bird images and specimens can also be observed in the economy of chicken farms. The meanings of storing and stockpiling in the world of birdwatchers can thus be understood in the light of expert analyses on livestock.

LIVESTOCK REVOLUTION AND ANIMAL LIBERATION

Microbiologists argue that when flu viruses emerge in wild birds, they are amplified in domestic poultry before spreading to humans. If wild birds are considered the animal reservoir of influenza, the dramatic amplification in domestic poultry, raising potential risks of pandemics, is qualified as a "livestock revolution."[48] Etymologically, this notion describes the process through which living diversity is transformed into a standardized stock that can be valorized and exchanged. Agriculture experts have used this term to describe the process of confinement, concentration, and integration that has turned small poultry farms into big factories after the Second World War. Such a process, starting in U.S. farms, has been so successful for the poultry industry that it was extended to the rest of the world and of the food industry.[49] But geologists have recently argued that the domestication of the red junglefowl (*Gallus gallus*) in south China, between 10,000 and 7,000 years ago, could be used as a marker of a new geological epoch in which humans change their environment, called the Anthropocene.[50] Whether it is dated in 1945 A.D. or 10,000 B.C., the notion of "livestock revolution" suggests that the emergence of pandemics is caused not by the hazardous mutations of nature but by the anthropogenic transformation of the environment and that humans are therefore responsible for future pandemics. Incidentally, it is a phenomenon that started long ago in China but that, amplified by the U.S. industry, comes back in China as a terrifying specter. In the 1970s, the U.S. model of poultry breeding was developed in Asia through companies like Charoen Pokpand (CP), run by a Chinese businessman who started his activities in Thailand and Guangdong—a process criticized by geographer Mike Davis as producing "monsters at our door."[51]

The notion of livestock revolution has been popularized by the ornithologist and geographer Jared Diamond in his famous *Guns, Germs and Steels*, published in 1997, the very year of the emergence of H5N1 in Hong Kong. In a chapter entitled "The Lethal Gift of Livestock," Diamond suggests that the shift from hunter-gatherer societies to pastoral societies increased the proximity between humans and animals, thus creating opportunities for pathogens to spread. When pastoralists gave food and care to their flock, they received animal products as a counter-gift but also viruses as a poisoned gift, to which they became immunized. Diamond thus explains the collapse of the Amerindian population by the contact with germs carried by European soldiers and their animals. Following this argument, the dramatic increase of domestic animals would have created a new form of alienation between humans and animals: domestic animals have become as foreign to modern humans as they were to Amerindians, and that is why they send pathogens. As the French-American physician René Dubos warned in 1968, nature has been transformed by humans in such a dramatic way that it "strikes back" by sending zoonotic pathogens.[52] Rather than taking this idea of "the vengeance of animals" as a driver of global human history, I want to see how it regulates relations between humans and birds in Chinese sentinel posts through techniques of preparedness.

In August 2005, in the aftermath of Hurricane Katrina and as H5N1 was moving from Asia to Europe, Michael Osterholm, an epidemiologist who had just been appointed by the secretary of the Department of Health and Human Services to the newly established National Science Advisory Board on Biosecurity, published an influential paper in the journal *Foreign Affairs* in which he justified the massive stockpiling of Tamiflu, flu vaccines, masks, and protective equipment launched by the Bush administration. His article ended with this figure, which was then often repeated in presentations of public health officials: "The population explosion in China and other Asian countries has created an incredible mixing vessel for the virus. Consider this sobering information: the most recent influenza pandemic, of 1968–69, emerged in China, when its population was 790 million; today it is 1.3 billion. In 1968, the number of pigs in China was 5.2 million; today it is 508 million. The number of poultry in China in 1968 was 12.3 million; today it is 13 billion."[53]

Given the reliability of statistics in China over this period, this figure is hard to confirm; but it shows the power of the narrative of livestock revolution as a factor of emerging infectious diseases. Following these figures, China is not to be dreaded for its cultural revolution but for its livestock revolution, which produces pathogens in 1997 more dangerous than Mao's ideas in 1968. However, this interpretation of the livestock revolution relies on a simplistic notion of causality that separates nature and culture, viruses and ideas. According to this narrative, as the number of livestock animals increases, the risk of pandemic viruses also rises. But it fails to describe how human representation and selection of animals as livestock affect its potential as a pandemic threat. Rather than discuss the production of these statistical figures, I will engage ethnographically the questions raised by the livestock revolution. If animals enter in contact with humans through gift-giving relationships, how does the treatment of animals as livestock transform gift into poison, and how does it affect relations between humans and birds, and among humans themselves?

The genealogy of the notion of livestock has been traced by Sarah Franklin in her book on the cloned sheep Dolly and the sheep-raising industry in the United Kingdom. "Livestock" can be described as one of the most important forms of biovalue, since it appears literally as the transformation of living beings into capital—shares that can be accumulated, circulated, and exchanged (as in the "stock market"). But one of the first meanings of "stock," Franklin notes, is the notion of stem or trunk, which is linked to a paradigm of kin and descent. "Livestock itself turns out to be a hybrid mixture with a distinctive pedigree. In the terms are combined the meanings of stock as tool (tool part) and fund (or capital) with the earlier senses of race or breed."[54] A cloned animal such as Dolly is thus at the crossroads between different modes of production and reproduction: standardized forms of measurement and more imaginary relations to life and death. In the same way, the idea of the livestock revolution as an animal reservoir for virus mutations combines two visions of living beings: standardized commodities that increasingly circulate around the globe and living beings who can revenge against humans by sending the "lethal gift" of pathogens.

The livestock revolution in south China transformed backyard poultry, considered a "sideline activity" of the household, into commodities that could be exchanged on the market. While "barefoot" veterinarians in Maoist China were not called when a duck or a chicken fell sick—by contrast with pigs or cows[55]—they became an integral part of poultry breeding after the reform movement in the 1980s, under the name of "avian medicine." In Hong Kong, Singapore, and Taiwan, the livestock revolution occurred in the 1950s, when Chinese immigrants were taught Western methods to raise chickens "scientifically."[56] The poultry industry in these three territories was triggered by the massive demand for Chinese breeds from the Chinese community in the United States, who could not buy these breeds in mainland China because of the U.S. embargo. While American experts developed an intensive poultry industry around Taichung, in central Taiwan, concentrating all the steps of poultry production, and while Singapore developed a poultry industry on the border with Malaysia to avoid dependence on importations, British authorities were suspected of raising chickens that were hatched in mainland China.

> [Many of the] ducks processed in Hong Kong and then exported to America came from eggs laid in China and brought to Hong Kong to hatch. Were the ducks from these eggs communist ducks or true-blue British ducks? The correspondence on the subject was voluminous before a solution was finally reached. Provided that an inspector was present when the duck was hatched, that he forthwith rubberstamped the duckling's foot, and that on reaching maturity a further marking was put on the duck, then the ducks might be slaughtered, dried and admitted into the United States.[57]

The livestock revolution thus involved separating all the steps of the production of life to map their specific risks. The division of labor in poultry production transformed the relation with a live animal into the selection of a breed that could be valorized and exchanged. In Hong Kong, the

Kadoorie brothers, two Jewish bankers coming from Iraq, built an experimental farm in the middle of the New Territories, where they developed sophisticated methods of selection, caging, feeding, insemination, and vaccination. In 1997, after the emergence of H5N1, the Kadoorie Farm could not sell its poultry products anymore and became a center for the conservation of local breeds, particularly the Wai Chow, the White Wai Chow, and the Guangzhou chicken. The Kadoorie Farm was proud to say that during the Cultural Revolution, Chinese pure breeds had disappeared from mainland China. It presented itself as a repository for local breeds, which could be sent to China to be raised anew. The visitor who enters the Kadoorie Farm today sees botanical gardens, wild birds, crocodiles, monkeys, and, preserved from the gaze of visitors, 2,000 chickens. A warning is sent to the visitor who might want to get close: "The Chicken Display House will be closed under further notice to ensure the chickens at the Kadoorie Farm and Botanical Gardens are protected from any possible outside contamination while bird flu concerns still exist in Hong Kong." The Kadoorie Farm has its own system of alert, more severe than that imposed by the government to other poultry farms, with three levels (vigilant, serious, urgent), because in case of an outbreak of bird flu in the surroundings of the farm, the cost of culling would not measure the value of the meat, but the genetic knowledge preserved by decades of selection.

It can be argued, therefore, that because of avian influenza, the Kadoorie Farm had to convert from a site of livestock production to a repository of biodiversity.[58] However, the same ethical dilemmas are transversal to both practices, as revealed by my interviews with Tam Yip Shing, the head of the breeding team, in charge of the 2,000 chickens and nine pigs in the farm. A passionate birdwatcher and plant scientist trained at Hong Kong University, he had wished to build his own farm; but the Environmental Impact Assessments were too stringent, and he accepted the job offer from the Kadoorie Farm. He told me that before 1997, the selection of the purest breed was a public ceremony, but that it became hidden after 1997 for safety reasons. Selection consisted in sexing the males from the females, ringing the females, preserving the males who had the highest value, and destroying the rest of the males. Shing told me, "In mainland China, people don't have the concept of species; for them, it is just meat." He contrasted the killing of one-day chicks for selection to the massive killing of poultry as a preventive measure against bird flu. "We use CO_2," he said. "This is not torture. For ten seconds they shake a lot, but after twenty seconds, it is silent. When they killed poultry at Cheung Sha Wan, the quantity of gas was not enough. Poultry died after a very long time. It was really torture. People watching on television felt distress."[59]

Despite his training as a birdwatcher, Shing's view of poultry farming may be described as pastoralist. He justified the killing of chickens by the selection of the pure breed, which is an ideal to be preserved above the lives of those beings the farmer had to take care of. For Shing, the opposition between livestock and sensitive animal was secondary to the maintenance of a Chinese essence that regulates the circulation of all living beings. Here we encounter an ethical dilemma that we have already seen in the chapter on simulations: the distinction between those who are made to live and those who are left to die is justified by the urgency to save the breed. The life of protected breeds is stocked just as vaccines are stocked for times of emergency: there is an implicit operation of triage in the selection of the pure breed.

However, the Kadoorie Farm reveals other modes of existence of birds in addition to livestock. Other cages opened to the public displayed the smuggled birds (mostly raptors and parrots) that the farm received from the border patrol. On Sundays, the staff released these birds publicly as a mobilization to advertise about the risks of religious practices of animal release. Practiced for a long time in Chinese societies by wealthy aristocrats or by Taoist priests, the liberation of animals (*fangsheng*, literally "let live") had developed in the last twenty years, accompanied by Buddhist movements as a consequence of the development of an urban middle class.[60] Ordinary citizens, most often under the guidance of a Buddhist monk, met in live animal markets and bought birds or fish that they released in a proximate area. In Hong Kong, dozens of dead birds were found in the natural parks where these releases were practiced. Most of those birds had died from the stress of being released in an unknown environment after being stuck in cages. But they were thus exposed to infectious diseases such as avian influenza.

As I mentioned in chapter 3, Hong Kong birdwatchers, among whom were members of the Kadoorie Farm such as Mike Kilburn, organized a press conference with the microbiologists of Hong Kong University in May 2007. They publicized a map showing that cases of H5N1 in wild birds had been found around the Bird Market of Kowloon rather than close the bird reserve of Mai Po. They also took part in a conference organized by a Taiwanese society for the protection of animals in Taipei in 2004 to investigate the origins and forms of bird release in south China, involving ornithologists, Buddhist authorities, animal rights specialists, and anthro-

FIGURE 6.3. The release of sparrows by Buddhist practitioners in a natural park in Hangzhou. Photo by Frédéric Keck, June 2009.

pologists. After this environmental science campaign, Buddhist associations in Hong Kong officially recommended stopping the release of birds. They put posters on Buddhist temples with birds flying after release turning into cadavers with the following comment: "This is not releasing life, but releasing death" (*bu shi fangsheng, shi fangsi*). They also suggested replacing birds with animals bought at the seafood market: fish, crabs, shells, frogs, tortoises.

In Hong Kong, I was able to join one of these Buddhist groups meeting every Saturday at the seafood market of Tuen Mun to buy and release animals. The dynamic leader of the group, Daniel Lo, was working in an insurance company, and most of the twenty people joining the group were middle-aged or retired people. The youngest members of the group told me they had gone through a personal quest through Christianity and Confucianism, but that they found comfort in this regular practice of *fangsheng*.

Most of the members of the group had met through websites, as pictures of bird release and appointments for the next meetings were posted on the Internet. They first gathered at the entrance of the market, gave money to Daniel Lo, and then walked with him along the shops as he chose which animals he would save from a promised death. I was struck to see Daniel buy frogs, then change his mind for shells, and the shopkeeper butchered in front of our eyes the frogs Daniel had just held in his hands. Daniel argued that there was so much suffering and bad *karma* in the market that it was impossible to release all animals, but that the release of some of them would increase the "merits" (*gongde*) in the world. Before the release, they sang Buddhist songs; after the release, they prayed for the souls of sick or dead humans and animals around them. Then they went to a vegetarian restaurant, where they shared pictures of the release and a DVD about animal welfare produced by animal protection societies or Buddhist groups. Sometimes, a Buddhist monk would join them in singing the prayers and give them a course in Buddhist cosmology over a vegetarian meal.

The practice of *fangsheng* was regulated by an economy of souls, which doubled and paralleled the economy of living bodies but didn't really challenge it. If a market can be considered a place where livestock is assessed and exchanged, then the Buddhist extracted some of these beings from the real market to introduce them into a virtual market where the images of their souls circulated—either in the counting of the merits produced by the release or by the sharing of images of the release. The Buddhist officials I met said they condemned *fangsheng* as a material practice because it increased zoonotic risks for humans and animals, but they argued that *fangsheng* was a spiritual practice of praying for the souls of those who suffered. They were proud to mention that in 1997, when the Hong Kong government killed all the live poultry on the territory, master Koh Kwong, head of the Buddhist Association, went to several points all over the territory to purify them with holy water. The Buddhist Association thus repeated at a spiritual level the sacrificial gesture operated by the government through actual killing of birds. The Buddhists didn't contest the support given by the government to the livestock revolution by practicing animal liberation as a way to give them rights:[61] they considered that eating animals introduced bad *karma* into the world, and should be redeemed by praying for the souls of animals. Yet the government and the Buddhists shared a view of animals as a stock that should be managed on a territory, either of bodies or souls.

By contrast, birdwatchers viewed birds following lines of flight or flyways. They spread a rumor that was difficult to confirm, but revealed a lot about their ontological differences with the Buddhists. According to birdwatchers, Buddhist officials were allied with the local mafia (famously known as "triades") to organize the sale of birds for release and captured birds after release to sell them again at the Bird Market, thus participating in a real circulation of livestock on the territory. Indeed, birds sold for release on the bird market were mostly local breeds like sparrows, sold at the price of 30HKD, by contrast with birds valued for the quality of their songs or their aptitudes for fight, often coming from southeast Asia or Mongolia. When they were released, the birds escaped slowly from the cages and could easily be captured and sent back to the market. Birds were not properly released in the wild but took part in a wider ecology of souls and viruses, in which they were valorized as signs of fortune and merit.

The ambivalence of bird release is a recurring topic in the Hong Kong film industry. At the end of *The Killer* (1989), John Woo releases a bird to signal the moment when the two opposed heroes are about to kill each other. At the beginning of *Three Kingdoms* (2008), director Daniel Lee uses a bird to signal that the imperial army is using infected human bodies as biological weapons. In *Sparrow* (2008), Johnnie To follows the movements of a young woman caught between two conflicting gangs with images of a bird trapped in a cage and finally released. The aesthetics of bird release translates the ancient views of Chinese aristocrats, who opened the cages of birds to thank them for the beauty of their song, in the frontiers of the Hong Kong territory, expressing the aspirations of Hong Kong citizens to escape an endangered territory and their fear of being trapped again in the Special Administrative Region under Chinese sovereignty.[62] A bird with flu is a sign that the gift-giving relationship between humans and birds has been violated: birds send viruses rather than songs, colors, and flesh because their movements have been restricted.

An interview I did in Taiwan reveals another view on the tensions of bird release. Wu Hung-Chu was the head of the Society of Environment and Animal Protection, based in the suburbs of Taipei and funded in 1923 under Japanese colonization. A former Buddhist monk, he had been trained in animal welfare in the United States and had organized the conference on animal release in 2004, during which Hong Kong and Taiwanese birdwatchers and Buddhist associations shared their views. As a strong supporter of a multicultural society, he was keen not to blame any local group for animal cruelty—neither the Buddhists for bird release nor the Aboriginals for animal hunting. His main target was the unregulated trade of wild and domestic birds that produced a high quantity of suffering and occasionally diseases. By contrast with the Hong Kong government, the Taiwanese government was not transparent in publicizing cases of bird flu on the territory because of the strength of the poultry industry lobby. Consequently, Wu Hung-Chu used the risks of bird release to produce awareness of the conditions of the livestock industry in Taiwan.

Wu Hung-Chu also allied with the U.S. association PETA (People for the Ethical Treatment of Animals) to denounce the release of pigeons for races in the Taiwan straits. According to PETA, more than a million pigeons are released every day from boats in the Taiwan straits by pigeon-racing clubs, and only 1 percent survive to come back to their shelter.[63] Huge amounts of money (an estimated 2 billion USD) are spent in bets on those who will return, welcomed as heroes if they are first, killed if they are too slow. But even those who are considered winners in one race have to go through seven races in one year before being retired. The release is thus the fictional cover for an extremely severe form of selection in pigeon-breeding, a central activity in Taiwan.

Pigeons released from boats in the Taiwan straits can be compared to the black-faced spoonbills released from Tainan and mentioned in chapter 4 as two figures of the Taiwanese dream to return safely to mainland China.[64] But while black-faced spoonbills are followed by birdwatchers through GPS along East Asian flyways, pigeons are trained to return to Taiwan and registered for their value on the pigeon-race market. While black-faced spoonbills are stored in bird websites for the protection of the environment, following a hunting technique, pigeons are part of a livestock market that produces value through selection and destruction, following a pastoral rationality. Releasing pigeons in the Chinese straits is neither a ritual nor a simulation but a gambling practice through which birds compete to cross the sea between Taiwan and China. Without any compassion for their fate, humans identify with birds as soldiers engaged in a sacrifice between two nations in a potential war. The reverse identifications and complex variations that we have seen in simulations have been replaced by a repetitive play. When birds fail to be perceived as sentinels, they return as ghosts of living beings wrongly transformed into dead specimens.[65]

The analysis of different forms of bird breeding in Taiwan and Hong Kong thus confirms the general argument of this book: birdwatchers share hunting techniques of preparedness when faced with future uncertainties, while pigeon-race clubs, animal protection movements, religious groups, and public health administrations share the pastoral rationality of prevention. The livestock revolution, rather than a natural process causing the emergence of new pathogens, thus appears as an ontological shift from storage, as an equalitarian exchange with living beings, to stockpiling, as a mode of accumulation based on inequality. A tension in the relation between humans and wild birds, which we have seen in the discussion on the storage of viral strains and bird images by virus hunters and birdwatchers, thus becomes a contradiction in the management of livestock with pandemic potential by public health planners and global experts.

The situation of a birdwatcher producing books and databanks out of the encounter with wild birds can be compared to the work of microbiologists producing vaccines out of the mutations of viruses in live chickens. Both situations appear as unequal forms of gift-giving relationships: the bird gives something from itself (a picture, a sample) but doesn't receive an equivalent in return. The gift thus appears lethal if it is not properly returned, whereas if it is attenuated as an image or a vaccine, it provides immunity and protection. A situation of communication, in which birds and humans are considered equals in the role of sentinels, is turned into a situation of domination. This fundamental inequality in the human/animal relation is then extended in relations among humans. While storage is oriented toward equity, as a reflection of natural diversity available for all humans, stockpiling is an unequal form of possession, raising issues of distribution and sovereignty. The distinction between storage and stockpiling allowed me to study ethnographically different regimes of biovalue in the accumulation of viral samples and bird specimens in Chinese sentinel posts. But it also corresponds to a more fundamental distinction between cynegetic and pastoralist modes of production, considered techniques to anticipate the future in the management of human-animal relationships. I now turn to the anthropological debate on hunter-gatherers to clarify this distinction and its anthropological significance.

HUNTER-GATHERERS AND VIRAL TRANSFORMATIONS

In the 1970s, when microbiologists started to discuss the causes of emerging infectious diseases, the modes of production and accumulation of hunter-gatherers were at the heart of debates in social anthropology. While the notion of animal reservoir was used by biologists as a cause for the emergence of a new virus in a way that could not be calculated but only imagined retrospectively, the concept of mode of production describes a web of relations between humans and animals in which local tensions and inequalities produce historical shifts. At stake in these discussions within post-Pasteurian microbiology and post-Marxist anthropology were the following questions: how to depart from an evolutionist model of causality as a development from simple to complex phenomena, to think about a multi-leveled causality entangling viruses and bacteria, birds and pigs, ghosts and kins, national states and migrating populations? In the final section of this chapter, I want to reconsider these debates in the light of contemporary concerns for biosecurity. This discussion should shed light on the distinction between storage and stockpiling, which, I argue, could be considered a criterion to separate a hunting rationality of preparedness and a pastoralist rationality of prevention as two ways to manage the uncertainties of animal diseases.

Anthropologist Alain Testart has argued that storage was the first step toward sedentarism in hunter-gatherer societies because it provides the primary form of accumulation. Testart makes a distinction between two kinds of economy: hunter-gatherers who have immediate consumption, such as Australian Aboriginals, and hunter-gatherers who store the results of their collections to face natural disasters, such as North-West Coast Amerindians. Storage, he argues, is a result of overproduction, using techniques such as freezing, salting, dessication, fermentation, and construction of silos. If the first kinds of storage concerned plants and animals, they were in some cases later extended to raw materials, tools, or pottery.

Food storage, says Testart, brings about a "total change of mentality": the past is more privileged than the present to anticipate the future; nature is not seen as an ever-providing source but as a cause of disasters. It also introduces the first notion of surplus and thus the first social inequalities, without the organization of exploitation in social relationships. "In order to be prepared for any eventuality, Testart writes, "there will be a tendency to store a little more than the quantity usually needed."[66] If hunter-gatherers store goods, it means that the opposition between wild and domestic is not relevant in the description of food practices. Contesting the opposition made by Marxist archeologist Gordon Childe between hunter-gatherers and agriculturalists, Testart introduces a third category, "storing hunter-gatherers," to grasp a specific form of surplus on the threshold of domestication. Yet he joins Childe's evolutionist views when he describes all societies as deriving from a primitive form of totemism, which he describes as an intimacy between humans and animals regulated by kinship relations and expressed in the potentialities of the time of Dreaming.

Tim Ingold has criticized Testart's evolutionist assumptions, questioning the incompatibility between storage and nomadism. Hunter-gatherers, he argues, can move around a defined territory and yet return to the same points of subsistence.[67] For them, nature itself practices storage when plants develop their own forms of accumulation in a cyclical mode. There is no incompatibility either, according to Ingold, between storage and sharing: what is stored can be shared in the community without any recognition of property rights on living things or any form of "appropriation of nature," which for Ingold marks the threshold of domestication. Ingold proposes therefore a distinction between practical storage, which is a temporal delaying of the consumption of food resources, and social storage, which involves rights of property and obligations of reciprocity. Social storage subordinates different forms of kinship between elements in stock to the appropriation of nature by society, while practical storage follows the movements of natural cycles. In practical storage, knowledge about the environment (such as images of animal spirits) is conserved through artifacts carried by hunters in such a way that the accumulation of resources doesn't tame the uncertainties of the hunt.[68] Ingold's effort to distinguish hunting practices from concepts of property thus led him to describe the "life-world" of hunting societies phenomenologically, prior to its deformation by "the appropriation of nature."[69]

Tim Ingold's views on hunter-gatherers go furthest, among Marxist anthropologists, in asserting an ontological difference between hunters and pastoralists. One of his main interlocutors is Marshall Sahlins, who argued that hunting societies combine in a contradictory way underproduction (of resources) and accumulation (through kinship). Sahlins famously showed that the "domestic mode of production" is sufficient in itself, but that when it intensifies, its inner contradictions transform it into other modes of production.[70] Ingold describes their disagreement as follows: "Whereas for Sahlins, 'the inherent cleavages of the domestic modes of production are mystified by an uncritical ideology of reciprocity,' we contend that generalized reciprocity is an inherent property of the mode of production, mystified through the imposition of a concept of private ownership that renders obligatory sharing an enlightened generosity."[71] For Ingold, contrary to the evolutionist views of Testart and Sahlins, the transformation from hunting to pastoralist societies is not the necessary result of inner contradictions, but the unpredictable effect of a catastrophe that occurred within a fully coherent ontology. In Ingold's view, the event is exterior to the structure of reciprocity, while for Sahlins and Testart, it is carved in its internal contradictions.

This discussion on structure and event particularly sheds light on the perception of epizootics among hunters and pastoralists. Comparing milk pastoralism among the Nuer described by Evans-Pritchard with meat pastoralism among the Tungus observed by Shirokogoroff, Ingold notes that in both conceptions, "a man who lends a cow to a kinsman when he has animals to spare is entitled to a return of equivalent or greater value for the activation of a great many claims and counterclaims, the aggregate result of which will be a flow of animals from those who have suffered least to those worst affected."[72] The Nuer and the Tungus kill their domestic stock either as victims of sacrifice in nuptial and shamanistic ceremonies or as victims of diseases during epizootics. The loss of an animal in these two occasions is experienced as an act of God. For a pastoralist breeder, therefore, according to Ingold, epizootics take part in a circulation of livestock regulated by sacrifice and not in the maximization of value as in capitalist ranging. "His reason for accumulating stock lies not in a desire to increase yields beyond a fixed domestic target, but in the need to provide his household with some security against environmental fluctuations, given a system of productive relations which places the burden of the future on his own shoulders."[73] In hunting societies, by contrast, following Ingold, epizootics are perceived as a sign that the sharing of hunted animals has not been made respectfully, and interpreted as a revenge of the Master of animal spirits, who has responsibility for the distribution of living beings.

Even if it is inspired by the phenomenological analyses of Ingold, the method of this book is closer to Sahlins's structuralism. If the distinction between hunters and pastoralists led me to separate birdwatchers and virus hunters on one side from Buddhist officials and public health planners on the other, I don't oppose storage and stockpiling as reciprocity and hierarchy, because I show that both techniques of accumulation are embedded in gift-giving relations between humans and animals that contain tensions, contradictions, and inequalities. The problem I raised with animal diseases is how viruses reveal these contradictions and how modes of accumulation of viruses stabilize them. If a virus appears, following Diamond's phrase, as a "lethal gift," the problem is how the information carried by the virus, instead of destroying the economic system in which it appears, becomes part of its production of value. Following viruses allows anthropologists to see how gradual mutations in animal reservoirs—between the temperatures at which viruses are conserved or the species in which they transmit—produce transformations in the modes of production when they involve kinship relations and ghostly apparitions.

Paul Rabinow was one of the first to introduce in the anthropological debates about biotechnologies in the 1980s the questions that had been raised among Marxist anthropologists a decade before, particularly around the role of scarcity in the accumulation of biovalue. For him, the revolution of biotechnologies—such as the Polymerase Chain Reaction invented by Kary Mullis in California in the 1980s—transformed a scarcity of gene information into an abundance of biological material.[74] Rabinow forged the notion of biosociality to describe the way collective movements are gathered around a piece of information—for instance, about the sequence of the gene for muscular dystrophy in the case of patients' groups helping geneticists to produce the first map of the human genome in France.[75] In a similar way, I have described the alliance of microbiologists and birdwatchers in the production of information about virus mutations in avian reservoirs to understand what they reveal of a vulnerable environment. At the end of *Making PCR*, Rabinow thus makes a striking reference to Lévi-Strauss's notion of bricolage, defined as "an extraneous movement" (*mouvement incident*),[76] in a discussion with Marshall Sahlins on the relation between structure and event. For Rabinow, an event doesn't reveal a hidden symbolic system, but problematizes an emerging set of forms that stabilizes into an assemblage or an apparatus.[77] While Rabinow started from the announcement of the first map of the human genome to study conflicting norms and forms in France as a "purgatorial space," I have taken the emergence of H5N1 in Hong Kong as an opportunity to study the conflicting norms and forms in Chinese sentinel posts. Biosociality and biosecurity have emerged as stable apparatuses out of these complex relations between structure and event, providing anthropologists with new cases to study the transformations of life forms and forms of life at the turn of the twenty-first century.

This discussion on structure and event leads me back, once again, from the 1970s to the Cold War. Lévi-Strauss's conception of the event was formulated in discussion with the philosophy of history of Jean-Paul Sartre, which provided human consciousness with a capacity to react to the experience of scarcity by a revolutionary creation of values. Lévi-Strauss criticized Sartre for ignoring the experience of societies that don't rely on the linear temporality of consciousness, but rather on a stock of knowledge of their environment. Bricolage is the use of a set of tools or materials that have been assembled randomly, without reference to any project: it is "the contingent result of all the occasions there have been to renew or enrich the stock."[78] Even if they are mobile and don't have property rights, "savage societies" (or hunter-gatherers) have a stock of knowledge and resources that allows them to interpret any event in a pregiven order of things. Lévi-Strauss dedicates long passages of *Savage Mind* to stocks of old names that are used for the integration of newborns or new diseases into the collective.[79] Some societies such as the Tiwi of the Melville islands "have an inordinate consumption of names": they put common names in a reserve where they are prohibited and become proper names ready-for-use after a delay.[80]

Lévi-Strauss's famous distinction between cold and hot societies was made during the Cold War, while Lévi-Strauss himself was concerned with questions of storage at the Musée de l'Homme, where he was a vice director. Hot societies conceive the future as the unfolding of a mechanism that is already running in the present, on the model of the steam engine or the oil motor: the future expands the past capacities of the machine. Cold societies conceive the future as a confirmation of an order in the past, on the model of the mechanical clock: they "freeze" time so that any event finds a place in a past set of relations. As Eduardo Kohn recently noted, "freezing" time in this sense doesn't mean resisting change but interpreting human history through a set of forms that connect humans and nonhumans. "What is 'cold' here is not exactly a bounded society. For the forms that confer on Amazonian society this 'cold' characteristic cross the many boundaries that exist both internal to and beyond human realms."[81] Freezing time, then, is a way to unscale human temporality and rescale it in the different sequences of forms that can be found in relations between humans, animals, and plants. Freezing doesn't reduce historical change, but enlarges the stock of relations in which it becomes meaningful.

This analysis by Lévi-Strauss sheds light on the distinction between storage and stockpiling I used for viruses in a contemporary logic of preparedness. Storage is the conservation of all kinds of tools and materials inasmuch as they can be used for different purposes. Its aesthetic value comes from its capacity of ordering through the language of forms. Stockpiling is the conservation of samples and vaccines for a pandemic, which orients the production of the pharmaceutical industry and makes its values. This resonates with the fact that stockpiling vaccines is practiced at 4°C, while storage of samples needs -80°C: the presence of adjuvants introduces a live element that triggers the reaction of the immune system, while samples can be conserved for their antigenic information. Storing viruses and stockpiling vaccines are techniques of preparedness that manage ambivalent information in relation to an uncertain future. While storage creates forms out of any type of scarcity, stockpiling simulates a scarcity to orient its production of forms. The values created by pharmaceutical industries when they benefit from pandemic preparedness, which may look like pathological forms of hoarding, should then be compared to the widest range of values accumulated in practices of storage, similar to those of collectors.

CONCLUSION

In this book, I have connected discussions about the anticipation of the future in the global circulation of biotechnologies with debates about relations between humans and animals in local contexts, thus combining an anthropology of biosecurity with an anthropology of hunter-gatherer societies, taking seriously the notion of virus hunters and the genealogy relating birdwatchers to wildfowl hunters. I have contrasted hunting techniques of preparedness and pastoral techniques of prevention as two ways to mitigate the uncertainties in relations between human and nonhuman animals. Rather than calculating risks through statistics and culling potentially sick animals, virus hunters and birdwatchers imagine the movements of birds through artifacts such as viral samples, computer software, databanks, tags, dummies, decoys . . . On the sides of the opposition between "make live and let die" and "make die and let live", which Michel Foucault strikingly used to operate a genealogy of pastoral power in Europe,[1] I have described how virus hunters and birdwatchers mimic animals and defer their deaths, taking ethnographic cases from Asian societies.

Although this book has not considered in ethnographic detail and does not rely upon direct ethnographic work among historical hunter-gatherers, such small-scale societies have depended on reading the signs of nature in a way that many, if not most, contemporary societies have lost in the drive for industrial food production. Claude Lévi-Strauss thought deeply about this loss in his diagnosis of mad cow disease under the provocative title "We Are All Cannibals."[2] In this book, I have expanded upon his analysis to reflect about what the anthropology of "hunters" reveals about the management of animal diseases. I have used *pastoralism* for industrialized-administrative ways of thinking (following Foucault's understanding of pastoral care) and *hunting* for intensive attention to the semiotics of signaling from the cellular and immune system level to that of ecologies and habitats. A third term, *collecting*, mediates between these two, shifting from centralized, limited access forms (the colonial model of the museum) to distributed, informaticized databanks managed for—in principle if not in practice—open access and democratic production and use (the postcolonial model). To follow Lévi-Strauss's provocation, I would argue that all humans can take the perspective of birds when reflecting on viral mutations usually managed by pastoral techniques and that our contemporary digital technologies, far from distancing us from birds and other species, help us build new relations with them in a virtual space.

This reflection, however, has remained firmly grounded in the three Chinese sentinel posts in which I have conducted fieldwork with different lengths of stay, benefiting from their relatively small scales to think about avian influenza in a Chinese perspective. While the colonial and postcolonial history of these territories fitted more in starting to conceptualize one of the three techniques of preparedness—sentinel in Hong Kong, simulation in Singapore, stockpiling in Taiwan—I have circulated among these three territories to display contrasts and transformations in the management of viral mutations coming from birds. In Hong Kong, sentinel chickens warn farmers of the presence of H5N1. In Singapore, simulations of culling chickens or evacuating patients provide realism to the threat of a bird flu pandemic after the experience of SARS. In Taiwan, stockpiling vaccines and antivirals is part of a Cold War strategy of preparing for a potential invasion from China while building a diplomacy of exchange with the rest of the world. These three territories compete and cooperate at the same time to be the best prepared for a bird flu pandemic, and yet they adapt these techniques to their populations of birds and humans.

The term "avian reservoir" in the title of this book has thus played on its association in tropical medicine with "Asian reservoirs," as if all diseases came from wild animals in Asia. To avoid this stereotype of a confused entanglement of humans and animals in a distant part of the world, I have showed that relations between humans and birds in these three territories are varied and diverse, since they are inscribed in recent and ancient histories and orient contemporary techniques of preparedness toward the global future. If there is something "Asian" in "avian reservoirs," it doesn't mean that viewing animals as sign-bearers is specific to this cultural area. The hunting perspective on avian preparedness is a modality of thought available for all humanity, but it is more often focused on concerns for terrorism, as in America, or for food safety, as in Europe, while in Asia it has been explored in its own terms without projecting techniques of prevention or precaution.

In a previous book,[3] I asked if a bird flu pandemic should be considered a myth in a critical sense that is often used when reflecting on other environmental threats such as climate change, nuclear radiations, endocrine disruptors, or species extinction. Borrowing from Lévi-Strauss's analysis of Amerindian myths and from the anthropology of hunter-gatherers that was built after him, I suggested that we consider this notion not in a skeptical but in a positive manner as describing dense and reversible relations between humans and animals in societies where the anthropocentric separations have not been cut. A bird flu pandemic is one of the "myths" that tell us most eloquently about our current relations with the environment in a world where poultry is increasingly raised for human consumption at high zoonotic risks and where wild birds are threatened with extinction by the destruction of their habitat. A bird flu pandemic is thus the reversal, as in a mirror, of a bird species extinction: an imaginary revenge of the birds against the way humans treat them. "Humans are infected with bird flu" is one of these paradoxical statements whose displacements produce imaginary frameworks and that anthropology takes as starting points for ethnographic investigations. Since birds can be considered enemies or allies in the fight against infectious diseases, these two opposite values instantiate a sense of shared vulnerabilities in different contexts.

I have described three techniques that shape this global imaginary: sentinels, simulations, and stockpiling. Sentinels actualize these opposite values at different levels where living beings communicate by signals. Simulations insert them in sequences of action so that the threat becomes real. Stockpiling produces value out of this distribution through accumulation and distribution. Through these three techniques, "myth" (as a reciprocal exchange of perspectives between human and nonhuman animals) becomes "ritual" (through play and performance) and "economy" (managing the accumulation of information that comes out of this exchange). This

TABLE C.1

CYNEGETIC TECHNIQUES	Sentinels	Simulation	Storage
PASTORAL TECHNIQUES	Sacrifice	Scenarios	Stockpiling
ETHNOGRAPHIC SITES	Hong Kong	Singapore	Taiwan
PHILOSOPHICAL PROBLEM	truth	reality	equity
ANTHROPOLOGICAL DOMAIN	myth	ritual	exchange

"becoming" is not an evolutionary development or a degenerative decline, and I don't tell a story of progress or loss. Rather, following Lévi-Strauss, I have conceived sentinels as displaying all the potentialities of relations between self and other at different levels and scales for reasons of prestige and almost as an aesthetic experience, while simulation and stockpiling restrict these potentialities for the sake of action and profit. I have thus showed how the aesthetic values of sentinels in communicating with birds are transformed into more pragmatic and financial values when inserted into sequences of action—or scenarios—and stockpiling repositories. At the same time, each of the three techniques of preparedness retains its potentialities of communication, which led me to contrast simulation and scenario—there is a potentiality of play in the simulation that is lost in the scenario—in the same way as I contrasted storage and stockpiling. I also maintained a distinction between sentinel and sacrifice to specify that the value of this mode of signaling is wider than the collective killing of a victim. These anthropological analyses allowed me to raise the philosophical problems of truth, reality, and equity.

Displaying the similarities between the work of virologists and birdwatchers in reference to hunter-gatherers, I wanted to contrast it with the work of public-health planners who rationalize the effects of sacrifice in a population and to distinguish the monitoring of bird flyways from the culling of avian reservoirs. This book thus draws a link between biosecurity concerns in the discussions on global health with environmental conservation, which was already at the heart of the ecology of infectious diseases in the 1960s but has become more developed in natural reserves and cultural museums. Improving biosecurity infrastructures means being attentive to the conditions of life of birds and those who take care of them and sharing with equity the valuable products that come out of this interaction. One of the challenges of a "One Health" approach of avian influenza is to develop the participation of all actors involved in the management of emerging pathogens, which compels experts to reframe concepts of causality relating humans, animals, and microbes.

I have argued in this book for a shift in the reflection on preparedness from the short temporality of emergencies to the long temporality of ecologies. While most genealogies of preparedness start with the emergence of H5N1 in 1997 in the global context of the war on terror or with the formation of civil defense in the U.S. in 1945 in the context of the fear of a thermonuclear war, I have developed three alternative genealogies on longer scales of temporality: first, the emergence of the H1N1 virus causing "Spanish flu" at the end of the First World War, giving rise to virology and epidemiology in the attempt to anticipate the next pandemic; second, the Industrial Revolution linking the colonial expansion of Europe to the rest of the world, giving rise to ornithology and other natural sciences under the framework of Darwinism; third, the encounter between the first European explorers and the New World depicted as an abundance of wildlife in the sixteenth century, giving rise to social anthropology as an attempt to encompass the most diverse forms of humanity. In these three scales of temporality, preparedness takes different meanings, as it doesn't involve preparing only for the next terrorist attack or the next natural disaster, but also for the destruction of the environments in which humans have coevolved with other animals and with microbes, using birds as sentinel species announcing the next extinction and bird artifacts as stocks where signatures of the past and signs of the future are accumulated.

I have therefore questioned my own position as a social anthropologist in this genealogy. Rather than criticizing preparedness through conceptions of the social lent to models of prevention, I have tried to describe preparedness from within and to understand its own resources for critical thinking. While prevention now appears as a dream of presenting all the information about diverse ecologies in a general framework where the past and the future follow the same patterns, particularly illustrated in postwar institutions such as WHO, IUCN, and ICOM, we may have to turn to the cynegetic practices of virus hunters to anticipate an unpredictable future by communicating with birds through databases in which their signs become meaningful.

Virus hunters appear in this book as critical actors, particularly when they ally with birdwatchers to criticize the excesses of pastoral power with its sacrificial techniques. Techniques of preparedness applied to avian influenza are themselves confronted with the possibility of failure or lure. What is a successful monitoring? When is an early warning signal perceived adequately? These questions are seriously raised in the complex sets of relations where birds and humans exchange their perspectives about the future, such as in the laboratories built under the Antigone consortium in Europe or in the associations launched by birdwatchers in Taiwan to fight against a construction project. Sentinels oscillate between overreaction and complacency because they are founded on an unstable relation between humans and birds. The ecology of infectious diseases has showed that viruses are not intentional entities aiming at killing humans, but signs that the equilibrium between species in an ecosystem has been disrupted. Humans, described by epidemiologists as the "dead-end" of chains of zoonotic transmission, tend to think that they are at the center of the ecosystem when they are only one of its actors. Sentinels are one of the ecological notions that decenter humans by showing their dependence on other species.

NOTES

INTRODUCTION

1. Osterholm, "Preparing for the Next Pandemic"; Davis, *The Monster at Our Door*; Greger, *Bird Flu*; Kilbourne, "Influenza Pandemics of the 20th Century"; Tambyah and Leung, *Bird Flu*; Sipress, *The Fatal Strain*.

2. Scoones, *Avian Influenza*.

3. Garrett, *The Coming Plague*; Osterhaus, "Catastrophes after Crossing Species Barriers"; Quammen, *Spillover*; Wallace, *Farming Human Pathogens*.

4. Most of the interviews were done in English; sometimes they were done in Chinese *putonghua* with translation; they were done fewer times in Chinese without translation.

5. Shortridge, Peiris, and Guan, "The Next Influenza Pandemic," 79.

6. Ethnographies of pandemic preparedness in China—Kleinman et al., "Avian and Pandemic Influenza"; MacPhail, *Viral Network*; Manson, *Infectious Change*—have relied on other experts than Shortridge, Guan, and Peiris and therefore didn't notice the specificity of what they call "doing preparedness at the avian level." For an approach of pandemic preparedness in Asia from political sciences, see Enemark, *Disease and Security*. For a review of the history of epidemic diseases and their management in East Asia, see Peckham, *Epidemics in Modern Asia*. For an ethnographic description of pandemic preparedness from the perspective of Singapore, see Fischer, "Biopolis," and Ong, *Fungible Life*.

7. This book is an attempt to follow the transformations of "birds with flu" from places where they are produced to spaces where they are consumed, describing their entanglements and disentanglements when they are captured by scientific experts, public health planners, and company businessmen. This method is parallel to what Anna L. Tsing did with "pinetrees with mushrooms" when she and her collaborators followed the transformations of matsutake from Japan to Oregon, Yunnan, and Finland. Immune cells as mediators of communication play a role in this book that can be compared to what Tsing describes about spores and nematodes. The concept of sentinel is therefore a contribution to an anthropology of the frictions of globalization, endeavoring to unscale the projects of capitalism or pastoralism; see Tsing, *Friction*, and *The Mushroom at the End of the World*.

8. On the stigmatization of human populations by the notion of animal reservoirs, see Lynteris, "Zoonotic Diagrams." On the shift in the conceptions of animal reservoirs with the turn to biosecurity and the anticipation of emerging infectious diseases, see Keck and Lynteris, "Zoonosis."

9. See Lévi-Strauss, *Tristes tropiques*, 163, and Keck, "Lévi-Strauss et l'Asie."

10. The term is borrowed by Paul Rabinow from Pierre Bourdieu: see Rabinow, *Anthropos Today*, 84–85.

11. I borrow this definition of concepts from Viveiros de Castro, *Cannibal Metaphysics*, and the taste for ontological distinctions from Descola, *Beyond Nature and Culture*.

12. I borrow this distinction from Andrew Lakoff: see Lakoff, *Unprepared*. On cynegetic power, see Chamayou, *Manhunts*.

CHAPTER ONE: CULLING, VACCINATING, AND MONITORING CONTAGIOUS ANIMALS

1. For a general overview of the history of veterinary practices and their contribution to public health—an approach now called "One Health"—see Bresalier, Cassiday, and Woods, "One Health in History."

2. Stirling and Scoones, "From Risk Assessment to Knowledge Mapping"; Catley, Alders, Wood, "Participatory Epidemiology"; Gottweiss, "Participation and the New Governance of Life."

3. Karsenti, *Politique de l'esprit*.

4. Becquemont and Mucchielli, *Le Cas Spencer*.

5. Spencer, *Study of Sociology*, 1–4.

6. Wilkinson, *Animals and Disease*; Fisher, "Cattle Plagues Past and Present."

7. Evans-Pritchard, *The Nuer*. Evans-Pritchard observed that the devastating effect of cattle plague led the Nuer to turn to fishing. The Nuer told him that cattle plague had entered their land fifty years before; see Spinage, *Cattle Plague*, 619–20.

8. Wilkinson, *Animals and Disease*, 169–71.

9. Woods, *A Manufactured Plague*.

10. Law and Mol, "Veterinary Realities."

11. Thomas, *Man and the Natural World*, 76.

12. In that sense, as a critic of the emerging welfare state, Spencer belongs to the genealogy of pastoralism proposed by Michel Foucault. The problem Spencer raises is how not to govern too much a social reality that is conceived as a natural set of regularities. Foucault takes eighteenth-century epizootics as a case that justifies the intervention of pastoralist rationality, since the pastor governs a flock of living beings by knowing about its singular determinations; see Foucault, "Omnes et Singulatim."

13. Spencer, *Study of Sociology*, 80–90.

14. With a more refined theory of mind, this is more or less the anthropological program of an epidemiology of representations launched by Dan Sperber in *Explaining Culture*.

15. Beidelman, *W. Robertson Smith and the Sociological Study of Religion*, 3.

16. While cattle plague and foot-and-mouth disease are epizootics, which cannot be transmitted from animals to humans, tuberculosis is a zoonosis that transmits from cows to humans and vice versa. Badgers are the animal reservoir in which the pathogen mutates; see Enticott, "Calculating Nature"; Jones, "Mapping a Zoonotic Disease."

17. Rosenkrantz, "The Trouble with Bovine Tuberculosis"; Worboys, *Spreading Germs*; Mendelsohn, "'Like All That Lives'"; Gradmann, "Robert Koch and the Invention of the Carrier State." In Germany, Emil Adolf von Behring supported the hypothesis of transmission of tuberculosis from cows to humans.

18. Frege, *Logical Investigations*, 121.

19. See Daston and Galison, *Objectivity*, 193.

20. Robertson Smith was a member of the Free Church of Scotland. During his studies in Edinburgh, he wrote an essay attacking Herbert Spencer's theory of the materiality of the soul. "The qualified evolutionism he did embrace was quite enough to make him heterodox to the elders of the church in which he had been raised, and they had him tried for heresy"; Stocking, *After Tylor*, 63–64. He traveled in Egypt, Palestine, and Syria in 1878, where he learned Arabic and hunted antelopes.

21. Robertson Smith (*The Religion of the Semites*, 154) makes a distinction between superstitious or magical precautions, inspired by the fear of the supernatural, and moral or religious precautions, prescribed by the frequenting of a holy place.

22. Robertson Smith, *Religion of the Semites*, 160–61.

23. Delaporte, "Contagion et infection."

24. Barnes, *The Making of a Social Disease*; Brydes, *Below the Magic Mountain*.

25. Quoted in Lynteris, "Skilled Natives, Inept Coolies," 309.

26. Lynteris, *The Ethnographic Plague*. Lynteris proposes the notion of epidemio-logic (based on Lévi-Strauss's mytho-logic) to explain that the logical contradictions between epidemiological narratives find their roots in the ontology of animal diseases.

27. See Pickering, *Durkheim's Sociology of Religion*.

28. See Durkheim, *Elementary Forms of Religious Life*.

29. See Carter, *The Rise of Causal Concepts of Disease: Case Histories*; Lukes, *Émile Durkheim*; Rawls, *Epistemology and Practice*.

30. Although it is not the purpose of this book, it would be interesting to compare the notion of asymmetry in Pasteurian microbiology and Durkheimian sociology. Pasteur's starting point is the role of chemical asymmetry in molecular biology, and Durkheim's is the role of inequality in social morphology. This comparison would shed light on the convergence between François Jacob and Claude Lévi-Strauss in the 1960s around the notion of structure, when genetics and linguistics revolutionize and confirm at the same time the works of the founding fathers of microbiology and sociology.

31. Durkheim, *Rules of Sociological Method*, 89. In 1894, when Durkheim published his *Rules of Sociological Method*, two students of Pasteur made major discoveries following his principles. Émile Roux discovered the first use of a horse serum to cure humans, and Yersin discovered the bacillus of bubonic plague in Hong Kong. Pasteur died one year after. This chronological parallel has never been stressed by Durkheim's commentators, who most often focus on his debate with British and Italian evolutionary biology.

32. See Latour, *The Pasteurization of France*.

33. See Bourdieu, Piet, and Stanziani, "Crise sanitaire et stabilisation du marché de la viande en France." These historians show that the tuberculosis and trichinosis outbreaks of the 1880s led to a redefinition of food safety in Great Britain, from a concern with poisoning as a moral issue (defining penalties for butchers who sold rotten meat) to an investigation of infection as a natural and social process (looking for the cause of an epizootic outbreak).

34. See Berdah, "La vaccination des bovidés contre la tuberculose en France."

35. Moulin, *L'aventure de la vaccination*; Bonah, *Histoire de l'expérimentation humaine en France*.

36. See Lévy-Bruhl, *Primitive Mentality*; Keck, *Lucien Lévy-Bruhl, entre philosophie et anthropologie*.

37. See Bergson, *Two Sources of Morality and Religion*, 145: "What the primitive man explains here by a 'supernatural' cause is not the physical effect, it is its *human significance*, it is its importance to man, and more especially to a particular man, the one who was crushed by the stone. There is nothing illogical, consequently nothing 'prelogical' or even anything which evinces an 'imperviousness to experience,' in the belief that a cause should be proportionate to its effect, that, once having admitted the crack in the rock, the di-

rection and force of the wind—purely physical things which take no account of humanity—there remains to be explained this fact, so momentous to us, the death of a man. The effect is contained pre-eminently in the cause, as the old philosophers used to put it; and if the effect has a considerable human significance, the cause must have at least an equal significance; it is in any case of the same order; it is an *intention*."

38. See Keck, "Bergson dans la société du risque"; "Assurance and Confidence in *The Two Sources of Morality and Religion*."

39. Lévi-Strauss, *Totemism*.

40. See Schwartz, *How the Cows Turned Mad*. The technical term for the disease was bovine spongiform encephalopathy, in reference to the sponge shape of the brains of the infected humans and animals.

41. Anderson, *The Collectors of Lost Souls*; Lindenbaum, *Kuru Sorcery*.

42. Lévi-Strauss, "La crise moderne de l'anthropologie," 14. On this vision of the "fault of the West," see Stoczkowski, *Anthropologies rédemptrices*, 253.

43. This is the conclusion of Robert Glasse's article published by Lévi-Strauss in *L'Homme* under the title "Cannibalisme et kuru chez les Foré de Nouvelle-Guinée."

44. In this article, Lévi-Strauss reacts to the news of the transmission of the Creutzfeldt-Jakob disease to young adults through the injection of hormones, which was a scandal in France in the context of the revelations of the transmission of HIV through blood transfusion. Rather than criticizing the public health administration, as journalists and officials did at the time, Lévi-Strauss notices the proximity between "ingestion and injection" to draw a parallel with cannibalism (there was yet no talk about "cannibal cows" at that time). He then argues with the anthropologist William Arens, who had contested the reality of cannibalistic practices among the Foré in Papua New Guinea.

45. On the use of Amazonian cosmologies to argue about the "denaturation" of animals in contemporary societies, see Erickson, "De l'acclimatation des concepts et des animaux, ou les tribulations d'idées américanistes en Europe."

46. Lévi-Strauss, *We Are All Cannibals*, 115.

47. See Derrida, "The Animal That Therefore I Am," 395. Derrida takes seriously the ontology of viruses transmitted from animals to humans. But he conceives the relation between living beings through the technologies of writing and reading rather than through relations of predation and commensability. "Neither animal nor nonanimal, organic or inorganic, living or dead, this potential invader is like a computer virus. It is lodged in a processor of writing, reading and interpretation"; Derrida, "The Animal That Therefore I Am," 407.

48. Descola, "Les avatars du principe de causalité."

49. Descola, *Beyond Nature and Culture*. From a perspective Descola calls naturalistic, we are led to ask how we can perceive cows as humans when

they are affected by common microbes; but from what he calls an "animist" perspective, microbes common to humans and animals are the real entities, and the separations made by humans to mitigate their threats are secondary constructions to stabilize these entities. In a third ontology Descola calls analogistic, mostly developed in Asia, Africa, and Central America, the proliferation of microbes in an environment is mitigated by a collective act of killing described by anthropologists as sacrifice. Descola criticizes Durkheim for describing the transformation from animism to analogism using data from totemic societies, which follow a different logic, as it attributes properties to common ancestral beings.

CHAPTER TWO: BIOSECURITY CONCERNS AND THE SURVEILLANCE OF ZOONOSES

1. The FP7 project "Antigone" was funded by the European Commission from 2011 to 2016 after bacteria resistant to antibiotics killed around fifty people in Germany, amounting to a health crisis in Europe. The bacteria were first thought to be present in Spanish cucumbers, but then the investigation showed that they had originated from cows and were transmitted through sprouts cultivated on Egyptian farms; see Keck, "Des virus émergents aux bactéries résistantes."

2. See Kuiken et al., "Host Species Barriers to Influenza Virus Infections"; Gortazar et al., "Crossing the Interspecies Barrier."

3. Calvert, "Systems Biology, Big Science and Grand Challenges."

4. See Biagioli and Galison, *Scientific Authorship*.

5. The names of influenza viruses are given according to the configuration of the hemagglutinin (H) and the neuraminidase (N) proteins that command the entrance and the release of the virus from the cell.

6. See Kolata, *Flu*, and WHO, "Influenza."

7. C. Hsu, "Critics: Airborne Flu Research Important, But Not for Vaccine Purposes," *Medical Daily*, February 8, 2012.

8. Zoe Butt, "Voracious Embrace." See also my article on Lêna Bùi's work: Keck, "Bird Flu: Are Viruses Still in the Air?"

9. See Porter, "Ferreting Things Out."

10. On the distinction between direct and indirect action in pastoral technologies of cultivation, see Ferret, "Towards an Anthropology of Action."

11. On the use of ferrets as models for the experimental study of influenza viruses, see Caduff, *The Pandemic Perhaps*, 45.

12. See Bennett, "The Malicious and the Uncertain." Bennett criticizes the way biologists and their ethical advisors oppose the "good intentions" of true scientists with the behavior of "those who would seek to do harm" in the rest of the world. Following Dewey, but also a theological tradition going back to Augustine, Bennett argues that scientists should rather reflect on

the capacities they develop while inventing new forms as they emerge in the laboratories.

13. See Collier, Lakoff, and Rabinow, "Biosecurity"; Hinchliffe and Bingham, "Securing Life"; Lentzos and Rose, "Governing Insecurity."

14. The moratorium was proposed by thirty-nine influenza experts in a letter published by *Science* and *Nature* on January 20, 2012, extended indefinitely in March 2012, then suspended by the same type of letter on January 23, 2013.

15. Lipsitch and Galvani, "Ethical Alternatives to Experiments with Novel Potential Pandemic Pathogens."

16. Wain-Hobson, "H5N1 Viral Engineering Dangers Will Not Go Away."

17. Lakoff, "The Risks of Preparedness," 457. "Rather than a conflict between science authorities and a fearful public, or between open inquiry and the demands of security, the controversy should be seen as a conflict among experts over different conceptualizations of an uncertain situation. As the controversy has unfolded, a fracture has become evident in the existing alliance among scientists and public health authorities around the threat of avian influenza."

18. Fouchier, Kawaoka, et al., "Gain-of-Function Experiments on H7N9."

19. Lipsitch and Galvani, "Ethical Alternatives to Experiments with Novel Potential Pandemic Pathogens," 535.

20. See MacPhail, *Viral Network*, 111: "The lure of scientific knowledge about deadly strains of influenza steered the shift of public health slightly off course."

21. See Jeffery K. Taubenberger et al., "Initial Genetic Characterization of the 1918 'Spanish' Influenza Virus"; Duncan, *Hunting the 1918 Flu*.

22. "Reverse genetics" can be defined as the reconstruction of a living being out of its genetic markers; see Napier, *The Age of Immunology*, 2; Caduff, *Pandemic Perhaps*, 108.

23. Palese, "Don't Censor Life-Saving Science."

24. Caduff, "The Semiotics of Security."

25. Caduff, *Pandemic Perhaps*, 140.

26. Caduff, "Pandemic Prophecy, or How to Have Faith in Reason," 302: "Precaution thus enables actors to commit a leap of faith. It allows them to have trust in a particular kind of future, even if there is no evidence that this future is likely to materialize."

27. See Drexler, *Secret Agents*, 18 ("The most menacing bioterrorist is Mother Nature herself"); Webby and Webster, "Are We Ready for Pandemic Influenza?," 1522 ("Nature's ongoing experiments with H5N1 Influenza in Asia and H7N7 in Europe may be the greatest bioterror threat of all"); Specter, "Nature's Bioterrorist."

28. Burnet, *Natural History of Infectious Diseases*; Anderson, "Natural Histories of Infectious Diseases."

29. Fassin and Pandolfi, *Contemporary States of Emergency*, 13.

30. See Domingo et al., "Viruses as Quasi-species."

31. See Russell, "The Potential for Respiratory Droplet Transmissible A/H5N1 Influenza Virus to Evolve in a Mammalian Host."

32. Williams, *Virus Hunters*; de Kruif, *Microbe Hunters*. See Caduff, *Pandemic Perhaps*, 52–53.

33. See Eyler, "De Kruif's Boast."

34. See Creager, *The Life of a Virus*.

35. McCormick and Fischer Hoch, *The Virus Hunters*; Gallo, *Virus Hunting*. See also Moulin, "Preface," in Perrey, *Un ethnologue chez les chasseurs de virus*, 16.

36. Lederberg, "Infectious History," quoted in Anderson, "Natural Histories of Infectious Diseases," 39.

37. Lederberg, "Infectious History," quoted in Anderson, "Natural Histories of Infectious Diseases," 40.

38. Wolfe, *The Viral Storm*, 9. On the controversial figure of Nathan Wolfe and his role in the implementation of techniques of preparedness in West Africa, see Lachenal, "Lessons in Medical Nihilism."

39. See Wolfe et al., "Bushmeat Hunting, Deforestation, and Prediction of Zoonoses Emergence"; Wolfe, Dunavan, and Diamond, "Origins of Major Human Infectious Diseases."

40. Wolfe, *Viral Storm*, 3.

41. Wolfe, *Viral Storm*, 38.

42. Wolfe, *Viral Storm*, 48.

43. Wolfe, *Viral Storm*, 28.

44. See Narat et al., "Rethinking Human-Nonhuman Primate Contact and Pathogenic Disease Spillover."

45. I borrow the expression *taking seriously* from the debate on "the ontological turn in anthropology," which involves considering ontological statements about the world as referring to the real rather than denoting a cultural set of symbols; see Carrithers et al., "Ontology Is Just Another Word for Culture"; Kelly, "Introduction: The Ontological Turn in French Philosophical Anthropology"; Keck, Regher, and Walentowicz, "Anthropologie: Le tournant ontologique en action."

46. See Viveiros de Castro, *From the Enemy's Point of View*; Viveiros de Castro, "Cosmological Deixis and Amerindian Perspectivism." Shamanism can also be asymmetrical and vertical, which explains its transformations into prophetism and pastoralism; but in its "pure" form, it is symmetrical and horizontal. In the nineteenth century, a pandemic prophecy was thus made by a shaman woman from the Desana of the Makuku: "The white people sent her a box containing a flag and other ornaments used by the caboclos for their saints' day festival. In it they put a curse that put a measles epidemic among her followers. After the epidemic, Maria announced the end of the world, a time when all sinners would be turned into animals with horns and be

eaten by jaguars and spirits"; Hugh-Jones, "Shamans, Prophets, Priests and Pastors," 60.

47. Kohn, *How Forests Think.*

48. Pedersen, *Not Quite Shamans*, 4–5.

49. See Stépanoff, "Devouring Perspectives"; *Chamanisme, rituel et cognition chez les Touvas.*

50. See Abraham, *Twenty-First-Century Plague.*

51. See Mark Honigsbaum, "Flying Dutchman Mans the Species Barrier," *Guardian*, May 26, 2005.

52. Osterhaus, "Catastrophes after Crossing Species Barriers."

53. Koch's postulates are rules to prove that a disease is caused by a pathogen. They require (1) that the pathogen must be found abundant in sick bodies and not in healthy bodies; (2) that it can be isolated and replicated in live culture; (3) that it causes disease when successfully transmitted to another organism; and (4) that it must be identical to the first pathogen when isolated from the second diseased body; see Gradmann, "A Spirit of Scientific Rigour."

54. See Drosten et al., "Identification of a Novel Coronavirus in Patients with Severe Acute Respiratory Syndrome."

55. See Drexler, Corman, and Drosten, "Ecology, Evolution and Classification of Bat Coronaviruses in the Aftermath of SARS."

56. See Linfa and Cowled, *Bats and Viruses*, 4. Hendra virus appeared in 1994 in Australia, Nipah virus in 1998–99 in Malaysia and Singapore and then in 2001–4 in Bangladesh, with high lethality for humans. It was discovered that bats carried these viruses asymptomatically. They are classified in the same family of viruses called Henipah.

57. King, "Security, Disease, Commerce," 767.

58. King, "Security, Disease, Commerce," 776.

59. See Webster, "William Graeme Laver: 1929–2008."

60. See Doherty, *Sentinel Chickens*, 89; Vagneron, "Surveiller et s'unir?"

61. Laver, "Influenza Virus Surface Glycoproteins H and N," 37.

62. Webby and Webster, "Are We Ready for Pandemic Influenza?," 1519.

63. Webster, "William Graeme Laver: 1929–2008," 217.

64. Griffiths, *Hunters and Collectors*, 12. See also MacKenzie, *The Empire of Nature.*

65. Indonesia and Papua New Guinea play a crucial role in this space between Asia and Australia: after being the sites of investigation for early naturalists such as Alfred Russel Wallace, they have been at the heart of the research on mad cow disease and bird flu: see Anderson, *The Collectors of Lost Souls,* and Lowe, "Viral Clouds."

CHAPTER THREE: GLOBAL HEALTH AND THE ECOLOGIES OF CONSERVATION

1. Lakoff, "Two Regimes of Global Health," 75.

2. Figuié, "Towards a Global Governance of Risks"; Hinchliffe, "More Than One World, More Than One Health"; Bresalier, Cassiday, and Woods, "One Health in History."

3. Vétérinaires sans Frontières was created in 1983 in Lyon to promote breeding in developing countries. In 2003, it merged with the Centre International de Coopération pour le Développement Agricole, created in 1997, to become Agronomes et Vétérinaires Sans Frontières (AVSF). The French organization has twenty-eight members, and eleven similar organizations have been created in other countries in Europe and Canada, gathered in the network VSF International (https://www.avsf.org/). GRAIN is an NGO created in France in 1982 to support genetic diversity in farms (https://www.grain.org/fr).

4. The term *One World, One Health* was coined by the Wildlife Conservation Society when it organized a series of meetings starting in 2004 on public health, conservation, and infectious diseases; see http://www.oneworldonehealth.org/.

5. Alpers, "The Museum as a Way of Seeing."

6. Bresalier, "Uses of a Pandemic."

7. Caduff, *The Pandemic Perhaps*, 42. See also Bresalier, "Neutralizing Flu."

8. Gaudillière, "Rockefeller Strategies for Scientific Medicine."

9. Hirst, "The Agglutination of Red Cells by Allantoic Fluid of Chick Embryos Infected with Influenza Virus."

10. WHO.IC/197, quoted in Aranzazu, "Le réseau de surveillance de la *grippe* de l'*OMS*," 400.

11. Caduff, "Anticipations of Biosecurity," 267.

12. Neustadt and Feinberg, *The Epidemic That Never Was*.

13. Boltanski and Esquerre, *Enrichissement*, 69: "The desire to possess some pieces in order to fill some gaps, defined by reference to an ideal totality, constitutes one of the main motivations for the behaviours among communities of collectors."

14. Strasser, "Experimenter's Museum," 62.

15. Strasser, "Experimenter's Museum," 79.

16. Strasser, "Experimenter's Museum," 91. "For Dayhoff, as for many naturalists in the past, collections and the items they contained were private property, and the collector was free to use them as commodities, gifts, or public goods. No item carried much value until it became part of a collection—that is, an element in a system designed for the preservation and production of knowledge."

17. See Sexton, *The Life of Sir Macfarlane Burnet*.

18. Burnet, *Natural History of Infectious Diseases*, 5. Israel Walton (1789–1863) was a Quaker poet settled in Swarthmore, Pennsylvania. Gilbert White (1720–1793) was the vicar of Selborne in Hampshire. His book, *Natural History and Antiquities of Selborne*, is considered the foundation of British ornithology.

19. Bargheer, *Moral Entanglements*, 46.

20. Le Roy, *Lettre sur les animaux*, 77.

21. Bargheer, *Moral Entanglements*, 10–11.

22. Bargheer, *Moral Entanglements*, 51–59.

23. Strivay, "Taxidermies."

24. Lewis, *Feathery Tribe*, 4.

25. Lewis, *Feathery Tribe*, 49.

26. Quoted in Lewis, *Feathery Tribe*, 48.

27. Lewis, *Feathery Tribe*, xii.

28. Quoted in Lewis, *Feathery Tribe*, 68.

29. Lewis, *Feathery Tribe*, 129.

30. Lewis, *Feathery Tribe*, 97.

31. Lewis, *Feathery Tribe*, 134.

32. Barrow, *Nature's Ghosts*, 152–53.

33. Quoted in Bargheer, *Moral Entanglements*, 119.

34. Bargheer, *Moral Entanglements*, 137.

35. See Ingrao, *The SS Dirlewanger Brigade*.

36. See Manceron, "What Is It Like to Be a Bird?"

37. See Adams, *Against Extinction*; Heise, "Lost Dogs, Last Birds, and Listed Species"; Sodikoff, *The Anthropology of Extinction*; Van Dooren, *Flight Ways*.

38. Moore, "Indicator Species," 3, quoted in Bargheer, *Moral Entanglements*, 368.

39. Moss, Interview, June 2006, quoted in Bargheer, *Moral Entanglements*, 189.

40. Findlen, *Possessing Nature*, 4, 9.

41. This is a quote of the first historian of the Royal Society, Thomas Sprat, in 1667, from Findlen, *Possessing Nature*, 400.

42. Yanni, *Nature's Museums*, 156.

43. Vidal and Dias, *Endangerment, Biodiversity and Culture*, 1.

44. Harrison, "World Heritage Listing and the Globalization of Endangerment Sensibility," 214.

45. Rabinow, Preface, in *Object Atlas*.

46. Gorgus, *Le magicien des vitrines*.

47. Laurière, *Paul Rivet*.

48. Malraux, *Le Musée imaginaire*.

49. Price, *Paris Primitive*.

50. Chiva, "Qu'est-ce qu'un musée des arts et traditions populaires?," 159.

51. Beltrame, "Un travail de Pénélope au musée"; Roustan, "Des clefs des réserves aux mots-clefs des bases de données."

52. Claude Lévi-Strauss conceived the Laboratoire d'anthropologie sociale of the Collège de France as a space where computers would help humans preserve massive stockpiles of data, under the conviction that after "savage societies" had disappeared, the files written about them were themselves vulnerable. He borrowed from Bert Kaplan the project of a microcard database, whose ambition was "to revolutionize the storage of social scientific data sets that were effectively, like the California Condor or the roseate spoonbill, facing extinction from forces of neglect or active harm (such as floods, fire and poor storage conditions)"; Lemov, "Anthropological Data in Danger," 97.

53. Appadurai, *The Social Life of Things*; Marcus and Myers, *The Traffic in Culture.*

54. Clifford, *The Predicament of Culture.*

55. See Brown and Kelly, "Material Proximities and Hotspots."

CHAPTER FOUR: SENTINELS AND EARLY WARNING SIGNALS

1. On the similarities and differences between contemporary sentinel devices and traditional divination, see Keck, "Ce virus est potentiellement pandémique."

2. See Rabinowitz et al., "Animals as Sentinels of Bioterrorism Agents." A sentinel is therefore an active agent in the fight against infectious diseases, in comparison with model animals, which remain passive; see Gramaglia, "Sentinel Organisms."

3. After an outbreak of Newcastle disease in southern California that led to the destruction of 12 million birds, the USDA launched a surveillance program with 37,000 sentinel chickens placed in 3,000 flocks for eight months. "The relative ease of placing and maintaining Specific Pathogen Free sentinel chickens, their susceptibility to many pathogens and the relative low cost makes them an attractive monitoring and surveillance tool for poultry diseases as well as certain human diseases"; McCluskey et al., "Use of Sentinel Chickens to Evaluate the Effectiveness of Cleaning and Disinfection Procedures in Non-Commercial Poultry Operations Infected with Exotic Newcastle Disease Virus," 296.

4. This device was also used in the 1960s and 1970s; see Doherty, *Sentinel Chickens*, 31–40. Doherty notes (103–12) that the malaria parasite, which is also borne by mosquitoes, was studied by Walter Ross on birds. Hawaiian birds have recently been devastated by malaria.

5. West Nile is an arthropod-borne virus (or arbovirus) transmitted to humans, birds, and horses by mosquitoes (there is no direct transmission from birds to humans). Originating from East Africa, it spread in the United States in the 2000s, starting in New York in 1999 (a strain previously identified in

Israel), and has killed 1,750 persons, mostly from encephalitis, with a lethality rate of 5 percent. Although birds are considered to be the natural reservoir of the virus, wild birds in the United States, particularly crows and jays, have died massively from West Nile due to their lack of immunity. The Audubon Society conducted a survey of corvids in the years 2004–6, concluding that the population of yellow-billed magpies had declined by 20 percent because of West Nile; see Doherty, *Sentinel Chickens*, 40–49, and Eidson et al., "Crow Death as a Sentinel Surveillance System for Westnile Virus in the Northern United States, 1999."

6. Yuen, "Clinical Features and Rapid Viral Diagnosis of Human Disease Associated with Avian Influenza A H5N1 Virus."

7. Woo, Lau, and Yuen, "Infectious Diseases Emerging from Chinese Wet-markets," 405.

8. Investigation Group on Epidemiological Study, *Epidemiology Report of the Highly Pathogenic Avian Influenza H5N1 Outbreak in December 2008 in a Chicken Farm in Ha Tsuen.*

9. Interview with Wang Yichuan, Ha Tsuen, February 15, 2009.

10. Interview with the Hong Kong Poultry Breeders Association, Yuen Long, December 16, 2008.

11. C. Chung, "'Town of Sadness' Pleads for Help," *Standard*, October 31, 2007.

12. Hanson, *Speaking of Epidemics in Chinese Medicine*. The infestation of grains was a major concern during the Great Famine in 1959; see Dikötter, *Mao's Great Famine*, 137.

13. Ben Striffler, who did fieldwork in a poultry slaughterhouse in New Mexico, notes that the migrant workers felt humiliated when their controller brought them Chicken McNuggets. "We're not going to eat this shit!," they exclaimed; Striffler, *Chicken*, 123.

14. Lévi-Strauss, *Les structures élémentaires de la parenté*, 67–71.

15. Porcher, *Eleveurs et animaux, réinventer le lien*.

16. Shortridge, Peiris, and Guan, "The Next Influenza Pandemic."

17. Elizabeth Etheridge (*Sentinel for Health*) recalls that the U.S. Centers for Disease Control were built in Atlanta because they were the "sentinel post" where the spread of yellow fever among mosquitoes could be observed and controlled in the south of the country. This mode of reasoning was applied by Kennedy Shortridge to Hong Kong.

18. Investigation Group on Epidemiological Study, *Epidemiology Report of the Highly Pathogenic Avian Influenza H5N1 Outbreak in December 2008 in a Chicken Farm in Ha Tsuen*, 12.

19. Kolata, *Flu*, 240.

20. See Doherty, *Sentinel Chickens*, 74, and Greger, *Bird Flu*.

21. M. Gladwell, "The Plague Year," *New Republic*, July 16, 1995.

22. Interview with H. C. Tsang, Headquarters of WHO in Geneva, June 21, 2007.

23. "Through imperial sacrifice, the emperor showed that things that seemed separate were in fact part of the same domain, i.e., his"; Zito, *Of Body and Brush*, 154.

24. See Manson, *Infectious Change*. On the conception of contagion in China, see Leung, "The Evolution of the Idea of Chuanran *Contagion* in Imperial China."

25. During the Cold War, the threat of the People's Liberation Army entering Hong Kong was perceived by the British officials as contributing to the distinct flavor of life in the colony. Ian Fleming wrote in 1963, "Hong Kong is a gay and splendid colony humming with vitality and progress. Knowing that six hundred and fifty million Communist Chinese are a few miles away across the frontier seems only to add zest to the excitement at all levels of life in the colony, and from the Governor down, if there is an underlying tension, there is certainly no dismay. Obviously China could take Hong Kong by a snap of its giant fingers, but China has shown no signs of wishing to do so. . . . Whatever the future holds, there is no sign that a sinister, doom-fraught countdown is in progress"; quoted in Miller and Miller, *Hong Kong*, 101–3.

26. In 1958, the city of Shanghai reported killing 1.3 million wild birds, the same number as poultry killed in Hong Kong in 1997; see Shapiro, *Mao's War against Nature*, 88.

27. Keck, "Live Poultry Markets and Avian Flu in Hong Kong,"

28. Liu Tik-Sang, personal communication. On the changes of traditional poultry production by biosecurity measures against bird flu in Hong Kong, see Liu, "Custom, Taste and Science."

29. Shortridge and Stuart-Harris, "An Influenza Epicentre?," 812.

30. Greger, *Bird Flu*, 35.

31. In the preface of the book written by Michael Greger, an animal pathologist from the United States, Shortridge says, "My mother's compelling stories about the devastating reaches of the pandemic have stayed with me since my earliest years. What started out as a spark of interest has led me to search the hows and whys of influenza pandemics through birds and mammals"; Greger, *Bird Flu*, xi.

32. When I asked him, "Why did you move to Hong Kong?," Shortridge answered, "This goes right back to my childhood and a series of events that took me to Hong Kong to 'get ahead of an influenza pandemic.' By 'getting ahead' was meant to try to detect the virus in humans at the start of a pandemic, make a vaccine, and distribute globally, as was then the case with the Hong Kong flu virus of 1968" (interview, Hong Kong, February 2, 2009).

33. Kennedy Shortridge, interview in Hong Kong, February 2, 2009.

34. See Powell, Watkins, Li, and Shortridge, "Outbreak of Equine Influenza among Horses in Hong Kong during 1992." The 1992 equine flu outbreak was also an opportunity for Shortridge to help Chinese authorities who had to deal with the same virus in Inner Mongolia in 1995.

35. Shortridge, "Avian Influenza Viruses in Hong Kong," 10.

36. Melinda Cooper defines preemption as "an aggressive counter-proliferation . . . to mobilize innovation in order to pre-empt its potential fall-out"; "Pre-empting Emergence," 121. Ben Anderson contrasts preemption with precaution: "Precaution is parasitic. It acts on processes that have an actual or possible existence prior to the intervention and does so on the basis of a determinate empirically apprehended threat. Preemption is different; it acts over threats that have not yet emerged as determinate threats, and so does not only halt or stop from a position outside. Its form of intervention is incitatory and it is justified on the basis of indeterminate potentiality"; "Preemption, Precaution, Preparedness," 14.

37. Sims et al., "Avian Influenza Outbreaks in Hong Kong, 1997–2002," *Avian Disease* 47, no. 3 (2003): 832–38.

38. See Leung and Bacon-Shone, *Hong Kong's Health System.*

39. See Greenfeld, *China Syndrome*, 211. "Peiris was the chief of human research in the Pandemic Preparedness group, while Guan Yi ran the animal side. After Malik Peiris announced that he had found a coronavirus, Guan Yi became obsessed with the idea of finding this virus's host species." A Japanese American journalist, Karl Taro Greenfeld, was editor of *Time Asia* in Hong Kong between 2002 and 2004 and interviewed many actors in the SARS crisis. He dramatizes the opposition between the soft-minded Peiris and the impetuous Guan and sometimes expresses stigmatizing views of the Chinese population. For a more academic perspective on the SARS protagonists, see Kleinmann and Watson, *SARS in China*; Abraham, *Twenty-First-Century Plague.*

40. Peiris et al., "Coronavirus as a Possible Cause of Severe Acute Respiratory Syndrome,"

41. See Peiris, "Japanese Encephalitis in Sri Lanka."

42. "During his research in animal influenzas, he liked to sometimes imagine himself as a virus. It was a trope he used in his talks, especially with his mainland counterparts. He often explained interspecies transmission by describing himself as a virus. . . . 'Oh, I like it here in my new house. I can take over this cell. I can reproduce. I'm a happy mutated virus'"; Greenfeld, *China Syndrome*, 212.

43. "When you have a virus or a pathogen that is adapted to a host, it comes into some form of equilibrium with the host. So in other words, if you take the human flu virus, it has actually developed a number of proteins that modulate the host response, that keeps the host response under control. Now when you have an avian virus, which is adapted to birds, and it comes to humans, it has never learnt this idea of modulating, or at least it was modulating the chicken response, but it is not modulating the human cell. So when you have those viruses that jump species, they don't know the rules, if you like (*laughter*), because they have not come to an equilibrium with the host. Of course, there are some people who say that if H5N1 adapts to human trans-

mission, it will automatically attenuate for the same reason. But I think in the long term it will—I mean the long term is ten years—but in ten years it could have killed many many millions of people. So I think it is not a good idea to just assume that adapting to human transmission it will automatically become less virulent"; interview with Malik Peiris at the Pasteur Centre, Hong Kong, October 7, 2007.

44. See Guan et al., "Isolation and Characterization of Viruses Related to the SARS Coronavirus from Animals in Southern China"; Shi, "A Review of Studies on Animal Reservoirs of the SARS Coronavirus."

45. Greenfeld, *China Syndrome*, 274–308. Jian Yanyong, physician at the Beijing's Chinese People's Liberation Army General Hospital no. 301, sent a letter to the Communist Party, asserting that there were more SARS patients in the capital city than the Chinese government had declared to WHO. Considered a "whistleblower," Jian Yanyong later told Greenfeld that the patients infected with SARS reminded him of students wounded at Tiananmen Square that he had treated in the same hospital in 1989. He was then put under house arrest.

46. Leung, "Efficacy of Chinese Medicine for SARS."

47. See Duara, "Hong Kong and the New Imperialism in East Asia 1941–1966." On the role of entrepôts in the globalization of commodities, see Roitman, "The Garrison-Entrepôt."

48. In 1957, Zhou Enlai declared, "Hong Kong should be converted to a useful port to our economy. . . . In the course of our socialist building, Hong Kong could become an operation base for us to establish overseas economic connections, and through Hong Kong we could attract foreign investment and foreign exchange"; quoted in Loh, *Underground Front*, 84. This strategy was implemented by Deng Xiaoping forty years later.

49. Carroll, *A Concise History of Hong Kong*, 160.

50. Bretelle-Establet, "French Medication in 19th and 20th Centuries China"; Peckham, "Matshed Laboratory."

51. Keck, "The Contaminated Milk Scandal."

52. Massive riots were organized by the Communist Party in Hong Kong in 1967 and ended just before the emergence of the H3N2 flu virus in July 1968; see Carroll, *A Concise History of Hong Kong*, 158–59, and Loh, *Underground Front*, 99–123. It can be said that in 1967, the Hong Kong government was preparing for the Cultural Revolution from China and fought against pandemic flu. Christine Loh writes, "In terms of preparedness, the colonial authorities benefited from police reforms introduced after the riots of October 1956"; *Underground Front*, 104. Inversely, in 1997, Hong Kong citizens perceived bird flu as indicating the threat of the arrival of the People's Liberation Army.

53. "Hong Kong's role as influenza sentinel is a little remarked-upon benefit of the "one country two systems" manifesto adopted by China upon the

colony's reuniting with the mainland"; Greenfeld, *China Syndrome*, 48. "The acronym (for SARS) was very similar to SAR, or 'Special Administrative Region,' the Chinese bureaucratic appellation for Hong Kong. Despite seeking to avoid any geographical stigmatization, the WHO had inadvertently done precisely that"; Greenfeld, *China Syndrome*, 219.

54. One World, One Health, http://www.oneworldonehealth.org/.

55. Whitney, "Domesticating Nature?"; Wilson, *Seeking Refuge*.

56. "Letter from Field Marshal Sir John Chapple," *HKBWS Bulletin* 207 (2008): 7.

57. The World Wildlife Fund was created in 1961 at IUCN headquarters in Switzerland, and Prince Philip was appointed president of the British branch. It has a global action in protection and conservation of the wildlife. Its Hong Kong branch was created in 1981. On the role of WWF in south China, see Hathaway, *Environmental Winds*.

58. On the analogies between monitoring and surveillance in British birdwatching, see Manceron, "Recording and Monitoring: Between Two Forms of Surveillance."

59. Masashi and Nagahisa, "In Memoriam: Elliott McClure 1910–1998." The final report was published by McClure under the title *Migration and Survival of the Birds of Asia*. But the observational data have been lost. See also Robin, *The Flight of the Emu*, 246–47.

60. Similar frictions between military global projects and environmental local concerns have been observed by Anna Tsing among naturalists in Indonesia: "Some nature lovers I spoke to had participated in army-sponsored training or competitions, and they recalled the discipline they learned there with considerable pride"; Tsing, *Friction*, 133. Tsing concludes that "environmentalism reproduced certain categories of New Order political culture—even as they challenged state policies"; *Friction*, 251.

61. Lam Chiu Ying, interview in Kowloon, December 8, 2008.

62. Lam Chiu Ying, "Thirty Years with the HKBWS," *HKBWS Bulletin* 207 (2008): 11. On the engagement of Hong Kong environmental movements such as Greenpeace, see Choy, *Ecologies of Comparison*.

63. Kilburn, "Railway Development Threatens Long Valley," 8. See also Allison, "An Object Lesson in Balancing Business and Nature in Hong Kong."

64. Simon Parry, "Closure Order on Mai Po Nature Reserve Is Lifted," *South China Morning Post*, March 18, 2004.

65. Letter from Ian Mckerchar sent to the Legislative Council of Hong Kong, followed by remarks from Geoff Carey: http://www.legco.gov.hk/yr05-06/english/panels/fseh/papers/fe0314cb2-1414-10-e.pdf.

66. Mike Kilburn, interview in Hong Kong Central, September 25, 2007. John Oxford is a professor of virology at the Royal London Hospital and was a highly visible expert at the time of the massive culling of birds for avian influenza in the United Kingdom.

67. AFCD, "Development of an Ecological Monitoring Programme for the Mai Po and Inner Deep Bay Ramsar Site."

68. Mike Kilburn, interview in Hong Kong Central, September 25, 2007.

69. Geoff Welsh, interview in Aberdeen, Hong Kong, July 15, 2012.

70. Severinghaus, Kang, and Alexander, *A Guide to the Birds of Taiwan*.

71. Lucia Liu Severinghaus, interview in Taipei, April 30, 2013.

72. Peter Chen, interview in Taichung, April 27, 2013.

73. See Weller, *Discovering Nature*. Hunter Eu got the first budget dedicated to conservation in any government agency and helped promote the first National Park Law in 1973. The Tourism Bureau is still the main sponsor of birdwatching activities in Taiwan today. Weller comments, "Here we can see the direct globalizing influence of the American wilderness ideal carried to Taiwan through the Park service"; *Discovering Nature*, 56. But national park management is not the only locus of ecological consciousness for birdwatchers. The concern for migratory birds was first addressed through the MAPS project, which Weller does not mention. Weller is right to say that "bird watching was one of the first signs of a changing popular conception of nature both in China and Taiwan" (*Discovering Nature*, 70), but he fails to see how the observation of nature was connected to a military project to control life—which may be explained by his lack of attention to birdwatchers in Hong Kong.

74. Lucia Liu Severinghaus, interview in Taipei, April 30, 2013.

75. Hsiao, "Environmental Movements in Taiwan," 36.

76. See Tang and Tang, "Local Governance and Environmental Conservation.

77. See Huang, "Saving Pillow Mountain, Taiwan."

78. See Veríssimo et al., "Birds as Tourism Flagship Species." For other wildlife species considered iconic, see Coggins, *The Tiger and the Pangolin*; Zhang and Barr, *Green Politics in Chinas*.

79. See Szonyi, *Cold War Island*. Lucia Liu explains the failure of Kinmen to turn into a sentinel of environmental threats by the feelings of its inhabitants that they have sacrificed too much during the Cold War. "Kinmen is very rich, it probably has too much money from their vineyards. They have a brewery, they make the most famous liquor for Taiwan, and the demand is so high. Nowadays zillions of Chinese go to Kinmen, they can just buy a ticket and go for a day tour, and they just buy everything they can bring back. But anyway, out of this huge amount of profit, there is little from the government. And they say: you've mistreated us, because we're the frontier, we're the battleground, we suffered. So the Taiwan government basically satisfied their demands. We are now part of that war, we didn't sacrifice, we benefited from their pain. So when we go there and say: 'don't develop this, look at your water supply, what are you going to do with fresh water?' they don't want to hear

it, they say 'we'll buy it from China.' They don't want to hear too much from those of us who didn't sacrifice"; interview in Taipei, April 30, 2013.

80. The distinction between self and non-self has been proposed by Frank Macfarlane Burnet as the central object for immunology, and then contested through a more complex analysis of the pathways of signaling among immune cells; see Moulin, *Le dernier langage de la médecine*; Martin, *Flexible Bodies*; Pradeu, *The Limits of the Self.*

81. On the metabolic process of freezing, nourishing, and destroying cells in a lab, see Landecker, *Culturing Life*, and "Food as Exposure.

82. Keck, "Feeding Sentinels.

83. For ethnographies of influenza viruses in the lab, see MacPhail, *Viral Network*, 53, and Caduff, *The Pandemic Perhaps*, 87.

84. Caduff, "The Semiotics of Security," 334.

85. Fox-Keller, *A Feeling for the Organism.*

86. Creager, *The Life of a Virus.*

87. The term *dendritic cell* was coined by Ralph Steinman in 1973, but synapse-shaped cells in the skin had been observed for the first time in 1868 by Paul Langerhans, who thought that they belonged to the nervous system. Steinman was the first to hypothesize that these cells played a central role in the immune system, which was formerly attributed to macrophages as "pathogen-eating" cells. He thus inaugurated a revolution in immunology that transformed the predator-prey relation between biological agents, conceived by Metchnikoff one century before, into a paradigm of communication of information. See Steinman and Cohn, "Identification of a Novel Cell Type in Peripheral Lymphoid Organs of Mice"; Banchereau and Steinman, "Dendritic Cells and the Control of Immunity." See also Anderson and Mackay, *Intolerant Bodies*, 125: "After 1973, sparsely distributed dendritic cells came to assume the major responsibility for antigen presentation, and hence, for activation of lymphocytes. It turns out that these previously obscure cells, scattered throughout the body, act as immunological sentinels, alert for pathogens and other intruders, ready to digest and exhibit antigens."

88. Kourilsky, *Le jeu du hasard et de la complexité*, 68, 106–9.

89. Kourilsky, *Le jeu du hasard et de la complexité*, 146–47.

90. Kourilsky, *Le jeu du hasard et de la complexité*, 204.

91. Napier, *The Age of Immunology*, 133.

92. Kourilsky, *Le jeu du hasard et de la complexité*, 174–75, 273.

93. Kourilsky, *Le jeu du hasard et de la complexité*, 280.

94. See Derrida, "Autoimmunity"; Anderson and Mackay, *Intolerant Bodies.*

95. Peiris et al., "The Role of Influenza Virus Gene Constellation and Viral Morphology on Cytokine Induction, Pathogenesis and Viral Virulence"; Cheung et al., "Induction of Proinflammatory Cytokines in Human Macrophages by Influenza A (H5N1) Viruses."

96. Salomon, Hoffmann, and Webster, "Inhibition of the Cytokine Response Does Not Protect against Lethal H5N1 Influenza Infection." Webster's team inhibited the gene for cytokines in mice and showed that they died of H5N1. But Peiris replied that they had died of encephalitis and not of respiratory disease, which is the cause of lethality for humans infected with H5N1.

97. Peiris and Porterfield, "Antibody-Mediated Enhancement of Flavivirus Replication in Macrophage-like Cell Lines"; Takada and Kawaoka, "Antibody-Dependent Enhancement of Viral Infection."

98. Peiris, Leung, and Nicholls, "Innate Immune Responses to Influenza A H5N1."

99. Mantovani et al., "Decoy Receptors"; Mantovani, Bonecchi, and Locati, "Tuning Inflammation and Immunity by Chemokine Sequestration."

100. Kourilsky, *Le jeu du hasard et de la complexité*, 21. Kourilsky makes a distinction between strategies of avoidance, such as mimicking cytokines to disrupt the defense reactions, and strategies of dissimulation, such as attaching silently to a cell before starting to infect it; *Le jeu du hasard et de la complexité*, 209.

101. Colborn, Dumanoski, and Myers, *Our Stolen Future*, 19: "This preoccupation with cancer had blinded her to the diversity of data she had collected. Moving beyond cancer proved to be the most important step in her journey, for as she looked at the material with new eyes, she gradually began to recognize important clues and follow where they led."

102. See Langston, *Toxic Bodies*; Wylie, "Hormone Mimics and Their Promise of Significant Otherness."

103. Colborn refers to a study done by Danish scientist Niels Skakkebaek on the decline of spermatozoids in human males, or Male Dysgenesis Syndrome. It is a review of publications about over 1,500 males in twenty countries, showing that the average number of spermatozoids produced by healthy men decreased by 45 percent between 1940 and 1990. Other studies have confirmed that result, particularly showing that spermatozoid speed was declining among younger men and linking it to an increase in testicle cancers among that population. These phenomena are explained by a growing exposure to estrogens in the maternal womb; other factors, such as smoking, alcohol, and sexual habits do not have the same effect on growth and affect all categories of males equally.

104. Another sentinel territory for the effects of endocrine disruptors is in French Martinique, where quantities of chemical products have been used in banana plantations beyond the doses authorized in the metropole, leading to the intoxication of a high proportion of the population; see Agard-Jones, "Bodies in the System."

105. See Wylie et al., "Inspiring Collaboration."

106. Zahavi and Zahavi, *Handicap Principle*, 4.

107. See Zahavi and Zahavi, *Handicap Principle*, 203.

108. See Zahavi, "Mate Selection."

109. Zahavi and Zahavi, *Handicap Principle*, 5. "If the babblers notify the raptor that they have seen it, both parties gain. The raptor moves on to another feeding ground, to try and surprise other prey; the babblers can resume their feeding. It makes sense for the babblers to signal to the raptor, and for the raptor to pay attention to their signal."

110. Zahavi and Zahavi, *Handicap Principle*, 40.

111. Zahavi and Zahavi, *Handicap Principle*, 194.

112. The philosopher Vinciane Despret, who followed Zahavi on the field at a time when the *Handicap Principle* was not yet published and his theory not yet accepted, defended him against criticisms for his reference to military war and sexual competition. For Despret, Zahavi reflects on his own scientific practice as "a creation of decoys." When sentinels send signals on their value in a situation of communication, they create beliefs in those who receive them, just as the scientist, faced with competitors who challenge his theory, creates beliefs through experimental decoys; Despret, *Naissance d'une théorie éthologique*, 161.

113. Philosopher Jean-Marie Schaeffer has thus used Zahavi's theory of costly signals to analyze the structure of aesthetic experience. For Schaeffer, Zahavi opens the way for a nonreductionist explanation of aesthetic experience in the evolutionary theory because he posits a structural homology, not a functional one, between sexual attraction and esthetic experience. For instance, bird dances are not determined by the sexual drive, but they express the same mode of production and reception of signals that can be observed in sexual behaviors. Schaeffer adds to Zahavi's theory a new element: the idea that a costly signal must be honest—that is, produce true information about the fitness of the sign-producer. "The central hypothesis of the theory of costly signals is that the cost or the benefit (for the one who sends the signal) of this type of handicap signal depends on the *real* qualities of the transmitter. The higher these qualities, the less costly is the signal for him; the lower they are, the higher is the cost. Insofar as the cost translates directly the qualities really possessed, a costly signal is a signal that cannot be simulated"; Schaeffer, *L'expérience esthétique*, 276. Schaeffer thus proposes a crucial distinction between simulation and fiction or between lure and decoy. While simulation lures or distracts by using the desires of the other, fiction plays on the intentions of the other and therefore can be called a decoy. In fiction, the transmitter declares clearly that a "costly signal" is sent, so that the receptor can adapt his/her behavior (*L'expérience esthétique*, 64). Schaeffer insists that signaling is both "honest" and "costly" because he wants to argue against sociological explanations of aesthetic experience that reduce it to an accumulation of prestige and leave aside the real value of individuals who produce or receive aesthetic objects (*L'expérience esthétique*, 286). Schaeffer targets Bourdieu's theory

of symbolic capital, following the analyses of Bliege and Smith, "Signaling Theory, Strategic Interaction and Symbolic Capital."

114. Zahavi and Zahavi, *Handicap Principle*, 229.

115. Zahavi refers to Thorstein Veblen's work on "conspicuous consumption," which explains waste by the survival of a "predatory phase of culture" within modernity; Zahavi and Zahavi, *Handicap Principle*, 160 and 227. Veblen himself referred to Franz Boas's observation on the "potlatch" of the Northwest Coast American Indians; Veblen, *The Theory of the Leisure Class*, 19. From Zahavi to Veblen to Boas, we are thus brought back to Claude Lévi-Strauss's theory of signs. Lévi-Strauss borrowed from structural linguistics the idea that there is a fundamental deficiency in signs, which explains the need to combine them with other signs to produce meaning. He applied this theory to kinship systems (with the prohibition of incest as the "negative rule" engendering all other rules) and to mythology (with the "mythem" as the meaningless element of a narrative that can combine with other "mythems" in an environment to produce meaning). While in his research on kinship he assumed that only humans could combine signs and create values, in his Mythologiques he recognized that the signals through which animals or cells communicate follow the same rules; the distinction between signs and signals thus became secondary. I have considered sentinels as mythems in Lévi-Strauss's sense: signals that produce values in the perception of a catastrophic discontinuity in the environment, whose accumulation could lead to the destruction of social life. The meaning of signs, according to Lévi-Strauss, constantly oscillates between two voids: their intrinsic lack of meaning and the absurdity of a general combination of signs without exteriority. See Lévi-Strauss, "The Lessons of Linguistics"; Keck, *Claude Lévi-Strauss: Une introduction*.

CHAPTER FIVE: SIMULATIONS AND REVERSE SCENARIOS

1. Sandrine Revet, who studied simulations of earthquakes in Peru, notices that this distinction is made in Spanish between *simulación* and *simulacro*: see Revet, "'A Small World.'"

2. An ethnography of the role of nonhumans in simulations of epidemics should mention the exercises preparing for zombie attacks organized in the United States in the last twenty years. These followed the success of popular books such as Max Brooks, *World War Z*, leading to a film starring Brad Pitt in 2013, where the scenario of pandemic emergence is linked to the contagious transformation of humans into zombies. The author of this book had previously published *The Zombie Survival Guide*, a genre that became popular with Joseph McCullough's *Zombies: A Hunter's Guide*. Admiral Ali Khan, head of the Office of Public Health Preparedness and Response at the U.S. CDC, wrote on the blog of the CDC: "The rise of zombies in pop culture has

given credence to the idea that a zombie apocalypse could happen. In such a scenario, zombies would take over entire countries, roaming city streets eating anything living that got in their way. The proliferation of this idea has led many people to wonder 'How do I prepare for a zombie apocalypse?' Well, we're here to answer that question for you, and hopefully share a few tips about preparing for *real* emergencies too!"; "Preparedness 101: Zombie Apocalypse," *Centers for Disease Control and Prevention*, http://blogs.cdc.gov/publichealthmatters/2011/05/preparedness-101-zombie-apocalypse/. I thank Maximilian Mehner, who kindly sent me the master's thesis he wrote on American Zombie Survival Camps for the ethnology department of the University of Marburg, "Zombie-Survival als Zeit-Phänomen."

3. Dhanasekaran et al., "Evolutionary Insights into the Ecology of Coronaviruses."

4. Smith et al., "Emergence and Predominance of an H5N1 Influenza Variant in China."

5. See Ong, *Asian Biotech*, and *Fungible Life*; Michael Fischer, "Biopolis."

6. See Linfa and Cowled, *Bats and Viruses*.

7. Smith et al., "Dating the Emergence of Pandemic Influenza Viruses." This warning could be compared to Fritz Bach's call for a moratorium on xenotransplantation from pigs, on the grounds that pigs are a reservoir of retroviruses, many of which remain unknown.

8. Donald McNeil, "Swine Flu May Have Come from Asia," *New York Times*, June 24, 2009.

9. Smith et al., "Origins and Evolutionary Genomics of the 2009 Swine-Origin H1N1 Influenza A Epidemic."

10. Mackenzie, "Bringing Sequences to Life."

11. Vijaykrishna Dhanasekaran, Interview, HKU–State Key Laboratory of Emerging Infectious Diseases, July 23, 2009.

12. Gavin Smith, Course on "Evolutionary Analysis of RNA Zoonotic Viruses," HKU–Pasteur Research Centre, July 22, 2009.

13. Justin Bahl, interview, Yung Shue Wan, July 15, 2009.

14. A recent change in the nomenclature of flu viruses has led to erasing the names of countries and provinces where they emerge in favor of their positions on the phylogenetic tree. Before this change of regulation, GenBank gave the name of the species and province where the virus had been found: for instance, A/Goose/Guangdong/1/96 H5N1 (or Gs/Gd) was linked with the H5N1 virus declared in Hong Kong in 1997 and considered to be its Chinese precursor. But other flu viruses found in China, such as the Fujian strain that spread around Asia and the Qinghai clade that spread in Europe were called "Clade 2.3.4" and "Clade 2.2." Reading phylogenetic trees thus entails a political knowledge of borders crossed by viruses, but this knowledge is coded in a language that displaces their political meaning into a biological anticipation. See Butler, "Politically Correct Names Given to Flu Viruses."

15. Bredekamp, *Darwins Korallen*; Helmreich, *Alien Ocean.*

16. Schüll, "The Gaming of Chance," 56.

17. Napier, *The Age of Immunology*, 2; Caduff, *The Pandemic Perhaps*, 105.

18. "Understanding the Flu," *Duke: Global Health Institute*, https://globalhealth.duke.edu/media/news/understanding-flu.

19. Lépinay, *Codes of Finance*, 80, 84–85.

20. See Peckham, "Economies of Contagion."

21. Hoong, *A Defining Moment*; James et al., "Public Health Measures Implemented during the SARS Outbreak in Singapore, 2003."

22. Andrew Zolli, "Learning from SARS," http://andrewzolli.com/learning-from-sars/.

23. UNSIC (United Nations System Influenza Coordination), *Simulation Exercises on Influenza Pandemic Responses in the Asia-Pacific Region* (2008), 56.

24. "AVA Holds Culling Exercise in Poultry Slaughterhouse," *Asia One*, July 17, 2013, http://news.asiaone.com/News/Latest+News/Singapore/Story/A1Story20130717-438043.html.

25. "Culling Exercise," *Straits Times*, Razor TV, July 17, 2013.

26. Hasnita A. Majid, "AVA Holds Culling Exercise to Test Readiness to Contain Bird Flu," *Channel NewsAsia*, January 10, 2008.

27. UNSIC, *Simulation Exercises on Influenza Pandemic Responses in the Asia-Pacific Region*, 63.

28. UNSIC, *Simulation Exercises on Influenza Pandemic Responses in the Asia-Pacific Region*, 18.

29. See V.-K. Nguyen, *The Republic of Therapy*; Redfield, *Life in Crisis.*

30. Interview with Ching Yong-Chung, Kowloon, December 15, 2011.

31. Anonymous interview by email after a questionnaire sent to the AMS, February 2009.

32. The opposition between active simulators and passive actors has also been described in Revet, "'Small World.'" In the simulation of an earthquake in Peru, actors bear signs of wounds such as red make-up and fake blood. On the simulation of a smallpox attack in the city of Albuquerque, New Mexico, Melanie Armstrong notes, "No amount of fake blood, however, can replicate the urgency that accompanies crisis, suggesting that these exercises are not valued because they provide new expertise on human behaviour, but because they rehearse forms of governing that are deemed to be important to managing a population perpetually in crisis"; Armstrong, "Rehearsing for the Plague."

33. On the way simulators explore potentialities in scenarios of disaster, see Samimian-Darash, "Practicing Uncertainty."

34. Barrow, *A Passion for Birds.*

35. Charvolin, Micoud, and Nyart, eds., *Des sciences citoyennes?*

36. See Wilson, *Seeking Refuge*, 76: "Waterfowl faced the threat of botu-

lism before the twentieth century, but the onslaught of water diversion and the growth of irrigated agriculture at the expense of wetlands made it much worse. The destruction of wetlands forced surviving populations of ducks and geese to congregate in small areas of remaining habitats. . . . Since waterfowl are gregarious by definition, epizootics of botulism could devastate birds on a given refuge very quickly. Migrating birds could also carry the toxin to other wetlands and refuges along the flyway."

37. Hong Kong birdwatchers often told me about their unlucky attempts at trapping black-faced spoonbills. In 1996, the AFCD used a rocket net fired by an explosive to catch black-faced spoonbills. In 2013, as they could not get the permit for the explosive, they had to use a whoosh net with an elastic band, which proved unsuccessful. Interview with Bena Smith, Hong Kong, October 18, 2015.

38. On the ethical dilemmas and technical precautions of satellite tracks imposed on wild animals for conservation purposes, see Benson, *Wired Wilderness*.

39. See Redfield, *Life in Crisis*, 164–65; Benson, *Wired Wilderness*.

40. Rollet, "Dimensions identitaire, sécuritaire et sociétale de la politique étrangère de Taiwan dans le domaine de la lutte contre les maladies infectieuses," 311.

41. See Cabestan and Vermander, *La Chine en quête de ses frontières*. In November 2000, General Zhang Wannian, vice president of the Chinese Military Central Commission, declared that a war in the Taiwan straits would necessarily occur in the next five years, reflecting the growing confidence of the Chinese PLA in its capacities of attack. The U.S. Department of Defense consequently provided its Taiwanese counterpart with techniques of simulation of military attack. A novel was published in English and Chinese in 2013 by Chuck Devore, vice president of the Texas Public Policy Foundation, and Steven Mosher under the title *China Attacks*.

42. "Mooncake Gambling Odds-On Festival Favourite," *China Daily*, September 28, 2004.

43. Szonyi, *Cold War Island*.

44. Zylberman, *Tempêtes microbiennes*, 90.

45. Galison, *Image and Logic*, 50.

46. Sismondo, "Models, Simulations and Their Objects"; Turkle, *Simulation and Its Discontents*.

47. Zylberman, *Tempêtes microbiennes*, 28, 153; Hamblin, *Arming Mother Nature*, 153–55; Galison, "The Future of Scenarios"; Lakoff, *Unprepared*, 23–24.

48. Kahn, *Thinking about the Unthinkable*, 143.

49. Ghamari-Tabrizi, *The Worlds of Herman Kahn*, 151.

50. Masco, *The Nuclear Borderlands*, 296.

51. Masco, *Nuclear Borderlands*, 305.

52. See Petryna, "The Origins of Extinction."

53. See Gusterson, *Nuclear Rites*, 160. "Just as, according to classical anthropological theory, the performance of rituals can alleviate anxiety and create a sense of power over, say, crops and diseases, so nuclear tests can in an analogous way create a space where participants are able to play with the issue of human mastery over weapons of mass destruction and symbolically resolve it. Since the stability that nuclear weapons are supposed to ensure—nuclear deterrence—exists so much in the realm of simulations, and since the reliability of deterrence involves the absence of a catastrophe more than the active, direct, positive experience of reliability, nuclear tests play a vital role in making the abstract real in scientists' lives."

54. Davis, *Stages of Emergency*, 4.

55. Ghamari-Tabrizi, *The Worlds of Herman Kahn*.

56. Davis, *Stages of Emergency*, 4.

57. Davis, *Stages of Emergency*, 51.

58. Davis, *Stages of Emergency*, 53.

59. Tomes, *The Gospel of Germs*.

60. Davis, *Stages of Emergency*, 41.

61. Lévi-Strauss, *Paroles données*, 149 (my translation).

62. Descola, *The Spears of Twilight*.

63. Houseman and Severi, *Naven or the Other Self*.

64. Houseman, "Dissimulation and Simulation as Modes of Religious Reflexivity," 82.

65. Houseman, "Dissimulation and Simulation as Modes of Religious Reflexivity," 87.

66. Filliozat, *Magie et médecine*, 79–80, quoted in Leiris, "La possession et ses aspects théâtraux chez les Ethiopiens du Gondar," 957.

67. Hamayon, *Why We Play*, 77. Roberte Hamayon is the author of a masterpiece in the study of Siberian shamanism: *La chasse à l'âme*.

68. Hamayon, *Why We Play*, 108.

69. Willerslev, *Soul Hunters*.

70. Hamayon, *Why We Play*, 204.

CHAPTER SIX: STORAGE AND STOCKPILING

1. Waldby, "Stem Cells, Tissue Cultures and the Production of Biovalue"; Rajan, *Biocapital*.

2. Bataille, *The Accursed Share*.

3. I thank Chang Chung-Ming for arranging this meeting.

4. "Six-Monthly Report on the Notification of the Presence of OIE-Listed Diseases," WAHIS Interface, http://www.oie.int/wahis_2/public/wahid.php/Reviewreport/semestrial/review?year=2012&semester=1&wild=0&country=TWN&this_country_code=TWN&detailed=1.

5. Lee, "Emergence and Evolution of Avian H5N2 Influenza Viruses in Chickens in Taiwan."

6. "Taiwan Finds H5N1 Virus in Birds Smuggled from China," *Medical Xpress*, http://medicalxpress.com/news/2012-07-taiwan-h5n1-virus-birds-smuggled.html.

7. "Taiwanese Woman Is the First Human to Be Sickened by H6N1 Bird Flu," *Los Angeles Times*, November 13, 2013; Wei et al., "Human Infection with Avian Influenza A H6N1 Virus.

8. Flu virus samples from birds are stored at 4°C for two weeks after collection, which is the time for virus identification and molecular sequencing, then at -20°C for rapid transport, and then at -80°C; Munster et al., "Practical Considerations for High-Throughput Influenza A Virus Surveillance Studies of Wild Birds by Use of Molecular Diagnostic Tests."

9. Landecker, *Culturing Life*, 227.

10. Kilbourne, "Influenza Pandemics."

11. See Neustadt and Feinberg, *The Epidemic That Never Was*. The vaccination against a new "swine flu" virus was suspended after more than 1,000 cases of Guillain-Barré syndrome were declared among those who received the vaccine.

12. Smith et al., "Emergence and Predominance of an H5N1 Influenza Variant in China." On the vaccine policy against bird flu in Vietnam, see Porter, "Bird Flu Biopower."

13. If a vaccinated chicken tests positive, it is impossible to know whether the antigens come from exposure to the flu virus or from the injection of the vaccine.

14. D. Silver, "Tiny Taiwan Preps for Worst; H7N9 Vaccine Plan in Place," *BioWorld Today* 24, no. 71 (2013); Chen, "Global Technology and Local Society."

15. Roy, *Taiwan*, 63.

16. Croddy, "China's Role in the Chemical and Biological Disarmament Regimes."

17. Keith Bradsher, "The Front Lines in the Battle against Avian Flu Are Running Short of Money," *New York Times*, October 9, 2015.

18. Rollet, "Dimensions identitaire, sécuritaire et sociétale de la politique étrangère de Taiwan dans le domaine de la lutte contre les maladies infectieuses (2000–2008)," 468, 533.

19. Rollet, "Dimensions identitaire, sécuritaire et sociétale de la politique étrangère de Taiwan dans le domaine de la lutte contre les maladies infectieuses," 664–66.

20. "Stockpiles of Anti-Virals in Europe," Wikileaks, https://wikileaks.org/gifiles/attach/96/96552_stockpiles of antivirals.doc.

21. "Detection of Human Swine Influenza Virus Resistant to Tamiflu," Government of Hong Kong, http://www.info.gov.hk/gia/general/200907/03

/P200907030213.htm. Resistant H1N1 strains to Tamiflu were also identified in Japan in 2009; resistant H7N9 strains were identified in China in 2013. Relenza could then be used as a complementary treatment.

22. Eric Tsang, "Hong Kong Running Out of Flu Drug as Lunar New Year Looms," *South China Morning Post*, February 9, 2015.

23. Minister of Health Siti Fadilah Supari declared in the journal *Nature*, "Samples shared become the property of the WHO collaborating centers in rich countries, where they are used to generate research papers, patents and to commercialize vaccines. But the developing countries that supply the samples do not share in these benefits. In the event of a pandemic, we also risk having no access to vaccines, or having to buy them at prices we cannot afford, despite the fact that the vaccines were developed using our samples"; Butler, "Q&A: Siti Fadilah Supari." See also Lowe, "Viral Clouds"; Hinterberger and Porter, "Genomic and Viral Sovereignty."

24. Vijaykrishna Dhanasekaran, interview in Hong Kong University, July 23, 2009.

25. On metaphors of "goldmine" or "black oil" to describe banks of viral strains, see MacPhail, *Viral Network*, 192.

26. Vijaykrishna Dhanasekaran, interview in Hong Kong University, July 23, 2009.

27. Fan, *British Naturalists in Qing China*, 135. A British consular official, Robert Swinhoe, was stationed in Xiamen and Ningbo between 1855 and 1875 and made observations in Hong Kong and Taiwan, where he was vice-consul between 1860 and 1866. His notes were published by the journal of the British Ornithological Union, *Ibis*, in 1861. Another British customs official involved in ornithology, John David Digues La Touche, stayed in Fujian between 1882 and 1921 and published a *Handbook of the Birds of East Asia* in 1925 and 1934.

28. Boutan, *Le Nuage et la vitrine*. Armand David, a French Lazarist priest, was sent to China as a missionary by Henri Milne-Edwards, administrator of the Museum National d'Histoire Naturelle, to collect zoological and botanical specimens. He explored Northeast and Southwest China between 1862 and 1874, sent more than 15,000 specimens to Paris, and discovered 60 new species. He published *Les Oiseaux de la Chine* with Émile Oustalet in 1877.

29. Rev. E. J. Hardy, *John Chinaman at Home: Sketches of Men, Manners and Things in China* (London: Fisher Unwin, 1907), quoted in Peckham, "Game of Empires," 213.

30. Fan, *British Naturalists in Qing China*, 156.

31. Fan, *British Naturalists in Qing China*, 22.

32. On the introduction of the Western concept of museum in Asia and its role in the emergence of nationalism, see Anderson, *Imagined Communities*.

33. Peckham, "Game of Empires," 218.

34. Ching, *Becoming Japanese*; Kikuchi, *Refracted Modernity*; Simon, *Sadyaq balae!*

35. *Yamashina Institute for Ornithology*, http://www.yamashina.or.jp/hp/english/index.html.

36. John Wu Sen-Hsiong, interview in Taichung, October 16, 2012.

37. Lucia Liu Severinghaus, interview in Taipei, April 30, 2013.

38. Dunlap, *In the Field, Among the Feathered*; Schaffner, *Binocular Vision*. While Dunlap situates bird books in a history of the environmental movement in the United States, Schaffner argues that the sanitized world represented in these guides misleads readers by omitting industrial landscapes and disconnects them from environmental degradation and its impact on bird populations. See also the classical analysis of Law and Lynch, "Lists, Field-Guides and the Organization of Seeing."

39. Trémon, "*Yingti/Ruanti* (Hardware/Software)," 138.

40. "Collections and Services: Natural History," *Hong Kong Museum of History*, http://hk.history.museum/en_US/web/mh/collections/collections/natural.html. It would be interesting to compare the fate of the natural history collections of the Hong Kong Museum of History in the context of the construction of the West Kowloon Cultural District (a massive project known as "M+") with the current reorganization of the collections of the Raffles Museum in Singapore. In 1823, Thomas Raffles, founder of Singapore, decided to create a museum as a repository for the natural and cultural specimens of the new British colony. The museum was opened in 1887, conserving also natural collections from the other founder of Singapore, William Farquhar. In 1960, the Raffles Library was moved out, and in 1965, after independence, the Raffles Museum became the National Museum of Singapore. The natural history collections were stored and conserved by scientists at the University of Singapore. In 1998, it was merged with the botany collections to form the Raffles Museum of Biodiversity Research. In 2014, the National University of Singapore raised 46 million dollars to erect a new building on the Clementi campus where the collections could be stored, researched, and displayed, thus becoming one of the largest natural collections of Southeast Asia with one million items; Lee Kong Chian Natural History Museum, https://lkcnhm.nus.edu.sg. After the poor conditions of storage of the first years at the university, the new Lee Kong Chian Natural History Museum will meet the highest standards of conservation, with three climate-controlled storage levels and cryogenic facilities; Barnard, "The Raffles Museum and the Fate of Natural History in Singapore," 184–211. By contrast with this high development of natural history collections, associations of birdwatching in Singapore are lightly organized. They don't have their own website but are part of the Nature Society of Singapore; "Birdwatching Hotspots in Singapore," https://www.nss.org.sg/wildbirdsingapore/Default.aspx. They meet irregularly to watch birds in nat-

ural parks such as Labrador Park or MacRitchie Reservoir and in members' flats to watch photographs. As for simulation, storage reveals that Hong Kong and Singapore have developed symmetrical potentialities of relations to birds that are much more entangled in Taiwan.

41. Mike Kilburn, interview in Hong Kong Central, September 25, 2007.

42. Mike Kilburn, interview in Hong Kong Central, July 8, 2011.

43. Ruy Barretto, interview in Hong Kong Central, July 14, 2011.

44. Barrow, *A Passion for Birds*; Moss, *A Bird in the Bush*.

45. On the databanks of naturalists, see Bowker, "Biodiversity Datadiversity."

46. See Charvolin, Micoud, and Nyhart, eds., *Des sciences citoyennes?*; Youatt, "Counting Species"; Maris and Béchet, "From Adaptive Management to Adjustive Management."

47. Lam Chiu Ying, interview in Hong Kong, December 8, 2008.

48. Delgado et al., *The Coming Livestock Revolution*.

49. Silbergeld, *Chickenizing Farms and Food*, 61.

50. Damien Carrington, "How the Domestic Chicken Rose to Define the Anthropocene," *Guardian*, August 31, 2016.

51. "Like the Green Revolution before it, the Livestock Revolution has favored corporate producers rather than peasants and family farmers"; Davis, *The Monster at Our Door*, 83.

52. Dubos, *Man, Medicine and Environment*.

53. Osterholm, "Preparing for the Next Pandemic," 35.

54. Franklin, *Dolly Mixtures*, 52.

55. Lyle Fearnley, personal communication, December 15, 2015. See also Fearnley, "Wild Goose Chase."

56. For Hong Kong, see Yeung, "Poultry Farming in Hong Kong." For Taiwan, see Lee, "Poultries in Taiwan." For Singapore, see Chou, "Agriculture and the End of Farming in Singapore." If, as we have seen in chapter 5, there is no poultry farm today in Singapore, and if the first British settlers failed to cultivate domesticated plants or animals in the island because of diseases (Turnbull, *A History of Singapore 1819–1975*, 44), the first government of Lee Kuan Yew insisted on developing a productive agriculture to sustain Singapore's autarchy after its failed alliance with Malaysia.

57. Grantham, *Via Ports*, 166. In 1960, during the Great Leap Forward, Hong Kong was the major destination of Chinese exports, and the British government often complained about the quality of these exports. See Dikötter, *Mao's Great Famine*, 110.

58. The work of the Kadoorie Farm can be compared to the mobilization of the Rare Breeds Survival Trust described by Donna Haraway: "RBST works against the premises and practices of factory farming on many levels, none of them reducible to keeping animals as museum specimens of a lost past or as wards in a permanent guardianship, in that utilitarian relations between

animals and people, including eating meat, are always defined as abuse. RBST maintains a database of breeds of poultry threatened with disappearance through industrial standardization; plans in advance how to protect rare-breed flocks from extermination by culling in bird flu and other epidemic disasters; supports husbandry conducive to whole-organism well-being of both animals and people; analyzes breeds for their most economical and productive uses, including new ones; and demands effective action for animal well-being in transport, slaughter, and marketing"; Haraway, *When Species Meet*, 273.

59. Interview with Tam Yip Shing, Kadoorie Farm, February 15, 2009.

60. Handlin Smith, "Liberating Animals in Ming-Qing China."

61. Singer, *Animal Liberation*.

62. Choy, *Ecologies of Comparison*.

63. "Tell the Taiwanese Government to Ban Deadly Pigeon Races Over the Ocean!," *Peta*, http://www.peta.org/action/action-alerts/first-ever-taiwan-raid-police-bust-pigeon-racers/. In August 2015, PETA announced that 129 persons from the Kaohsiung Zhong Zheng pigeon-racing club were charged for organizing these releases.

64. On the politics of managing living bodies migrating across the Taiwan Straits, see Friedmann, *Exceptional States*.

65. On the ghosts of soldiers in Kinmen island, situated between Taiwan and the Fujian coast of China, see Szonyi, *Cold War Island*, 181: "Ghost stories in Kinmen form a genre of the category of popular supernatural fictions in Taiwan today. These are stories of ghosts of dead soldiers wandering in Kinmen and haunting civilians, and ghosts of Kinmen villagers whose rest has been disturbed by the military constructions and who haunt the soldiers." Birds are often associated with ghosts in Kinmen because they appear on the beach, in the highly secured borderlands of the island. I thank Wang Xiyan for this observation.

66. Testart, "The Significance of Food Storage among Hunter-Gatherers," 527.

67. Ingold, "The Significance of Storage in Hunting Societies."

68. Charles Stépanoff, personal communication, October 7, 2015.

69. Ingold, *The Perception of the Environment*.

70. Sahlins, *Stone Age Economics*.

71. Ingold, *Hunters, Pastoralists and Ranchers*, 160.

72. Ingold, *Hunters, Pastoralists and Ranchers*, 170.

73. Ingold, *Hunters, Pastoralists and Ranchers*, 134.

74. Rabinow, *Making PCR*, 1.

75. Rabinow, "Artificiality and Enlightenment," in *Essays on the Anthropology of Reason*, 91–111.

76. Rabinow, *Making PCR*, 168.

77. Rabinow, *French DNA*, 180.

78. Lévi-Strauss, *Savage Mind*, 17.

79. Lévi-Strauss, *Savage Mind*, 164–65.
80. Lévi-Strauss, *Savage Mind*, 211–13.
81. Kohn, *How Forests Think*, 182.

CONCLUSION

1. Foucault, *Society Must Be Defended*.
2. Lévi-Strauss, *We Are All Cannibals*.
3. Keck, "Conclusion," in *Un monde grippé*.

BIBLIOGRAPHY

Abraham, Thomas. *Twenty-First-Century Plague: The Story of* SARS, *with a New Preface on Avian Flu.* Hong Kong: Hong Kong University Press, 2007.

Adams, Vincanne, Michelle Murphy, and Adele Clarke. "Anticipation: Technoscience, Life, Affect, Temporality." *Subjectivity* 28, no. 1 (2009): 248–65. https://doi.org/10.1057/sub.2009.18.

Adams, William B. *Against Extinction: The Story of Conservation.* London: Earthscan, 2004.

Agard-Jones, Vanessa. "Bodies in the System." *Small Axe: A Caribbean Journal of Criticism* 17, no. 3 (November 2013): 182–92. https://doi.org/10.1215/07990537-2378991.

Agriculture, Fisheries and Conservation Department, Hong Kong Government (AFCD). "Development of an Ecological Monitoring Programme for the Mai Po and Inner Deep Bay Ramsar Site." 2000. https://www.afcd.gov.hk/english/publications/publications_con/files/IssueNo14.pdf.

Allison, R. "An Object Lesson in Balancing Business and Nature in Hong Kong: Saving the Birds of Long Valley." In *Responsibility in World Business: Managing Harmful Side-Effects of Corporate Activity*, edited by Lene Bomann-Larsen and Oddny Wiggen, 121–37. New York: United Nations, 2004.

Alpers, Svetlana. "The Museum as a Way of Seeing." In *Exhibiting Cultures: The Poetics and Politics of Museum Display*, edited by Ivan Karp and Steven Lavine, 25–32. Washington, DC: Smithsonian, 1991.

Anderson, Ben. "Preemption, Precaution, Preparedness: Anticipatory Action and Future Geographies." *Progress in Human Geography* 34, no. 6 (April 2010): 777–98. https://doi.org/10.1177/0309132510362600.

Anderson, Benedict. *Imagined Communities: Reflections on the Origin and Spread of Nationalism.* London: Verso, 1983.

Anderson, Warwick. *The Collectors of Lost Souls: Kuru, Moral Peril, and the Creation of Value in Science.* Baltimore: Johns Hopkins University Press, 2008.

Anderson, Warwick. "Natural Histories of Infectious Diseases: Ecological Vision in Twentieth-Century Biomedical Sciences." *Osiris* 19 (2004): 39–61. https://www.jstor.org/stable/3655231.

Anderson, Warwick, and Ian R. Mackay. *Intolerant Bodies: A Short History of Autoimmunity.* Baltimore: Johns Hopkins University Press, 2014.

Appadurai, Arjun, ed. *The Social Life of Things: Commodities in Cultural Perspective.* Cambridge: Cambridge University Press, 1986.

Aranzazu, Anna I. "Le réseau de surveillance de la grippe de L'OMS: Circulation, innovation et santé publique." PhD diss., Université Paris 13, 2015.

Armstrong, Melanie. "Rehearsing for the Plague: Citizens, Security, and Simulation." *Canadian Review of American Studies* 42, no. 1 (spring 2012): 105–20. https://doi.org/10.3138/cras.42.1.105.

Banchereau, Jacques, and Ralph Steinman. "Dendritic Cells and the Control of Immunity." *Nature* 392 (March 1998): 245–52. https://www.nature.com/articles/32588.

Bargheer, Stefan. *Moral Entanglements: Conserving Birds in Great Britain and Germany.* Chicago: University of Chicago Press, 2018.

Barnard, Timothy B. 2014. "The Raffles Museum and the Fate of Natural History in Singapore." In *Nature Contained: Environmental Histories of Singapore*, edited by Timothy B. Barnard, 184–211. Singapore: NUS Press, 2014.

Barnes, David S. *The Making of a Social Disease: Tuberculosis in Nineteenth-Century France.* Berkeley: University of California Press, 1995.

Barrow, Mark V. *Nature's Ghosts: Confronting Extinction from the Age of Jefferson to the Age of Ecology*. Chicago: University of Chicago Press, 2009.

Barrow, Mark V. *A Passion for Birds: American Ornithology after Audubon.* Princeton, NJ: Princeton University Press, 1998.

Bataille, Georges. *The Accursed Share: An Essay on General Economy.* Translated by Robert Hurley. New York: Zone, [1949] 1988.

Becquemont, Dominique, and Laurent Mucchielli. *Le Cas Spencer: Religion, science et politique.* Paris: PUF, 1998.

Beidelman, Thomas O. *W. Robertson Smith and the Sociological Study of Religion.* Chicago: University of Chicago Press, 1974.

Beltrame, Tiziana N. "Un travail de Pénélope au musée: Décomposer et recomposer une base de données." *Revue d'anthropologie des connaissances* 6, no. 1 (2012): 217–37. http://doi.org/10.3917/rac.015.0255.
Bennett, Gaymon. "The Malicious and the Uncertain: Biosecurity, Self-Justification, and the Arts of Living." In *Modes of Uncertainty: Anthropological Cases*, edited by Paul Rabinow and Limor Samimian-Darash, 123–44. Chicago: University of Chicago Press, 2014.
Benson, Etienne. *Wired Wilderness: Technologies of Tracking and the Making of Modern Wildlife.* Baltimore: Johns Hopkins University Press, 2011.
Berdah, Delphine. "La vaccination des bovidés contre la tuberculose en France, 1921–1963: Entre modèle épistémique et alternative à l'abattage." *Revue d'Etudes en Agriculture et Environnement* 91, no. 4 (2010): 393–415.
Bergson, Henri. *Two Sources of Morality and Religion.* London: Macmillan, 1935.
Biagioli, Marco, and Peter Galison, eds. *Scientific Authorship: Credit and Intellectual Property in Science.* New York: Routledge, 2003.
Bliege, Rebecca B., and Eric Alden Smith. "Signaling Theory, Strategic Interaction and Symbolic Capital." *Current Anthropology* 46, no. 2 (April 2005): 225–48. https://doi.org/10.1086/427115.
Boltanski, Luc, and Arnaud Esquerre. *Enrichissement: Une critique de la marchandise.* Paris: Gallimard, 2016.
Bonah, Christian. *Histoire de l'expérimentation humaine en France: Discours et pratiques, 1900–1940.* Paris: Les Belles Lettres, 2007.
Bourdieu, Jérôme, Laetitia Piet, and Alessandro Stanziani. "Crise sanitaire et stabilisation du marché de la viande en France, XVIIIe-XXe siècles." *Revue d'histoire moderne et contemporaine,* numéro spécial, "Histoire de la sécurité alimentaire" 51, no. 3 (2004): 121–56. http://doi.org/10.3917/rhmc.513.0121.
Boutan, Emmanuel. *Le Nuage et la vitrine: Une vie de monsieur David.* Biarritz: Atlantica, 1993.
Bowker, Geoffrey C. "Biodiversity Datadiversity." *Social Studies of Science* 30, no. 5 (October 2000): 643–84. https://doi.org/10.1177/030631200030005001.
Bredekamp, Horst. *Darwins Korallen: Frühe Evolutionsmodelle und die Tradition der Naturgeschichte.* Berlin: Verlag Klaus Wagenbach, 2005.
Bresalier, Michael. "Neutralizing Flu: 'Immunological Devices' and the Making of a Virus Disease." In *Crafting Immunity*, edited by Pauline Mazumdar, Kenron Kroker, and Jennifer Keelan, 107–44. London: Ashgate, 2008.
Bresalier, Michael. "Uses of a Pandemic: Forging the Identities of Influenza and Virus Research in Interwar Britain." *Social History of Medicine* 25, no. 2 (2011): 400–424. http://doi.org/10.1093/shm/hkr162.
Bresalier, Michael, Angela Cassiday, and Abigail Woods. "One Health in His-

tory." In *One Health: The Theory and Practice of Integrated Health Approaches*, ed. Jakob Zinsstag et al., 1–15. Wallingsford, UK: CABI, 2015.
Bretelle-Establet, Florence. "French Medication in 19th and 20th Centuries China: Rejection or Compliance in Far South Treaty Ports, Concessions and Leased Territories." In *Twentieth-Century Colonialism and China: Localities, the Everyday, and the World*, edited by Bryna Goodman and David Goodman, 134–50. London: Routledge, 2012.
Brooks, Max. *World War Z: An Oral History of the Zombie War*. New York: Broadway, 2006.
Brooks, Max. *The Zombie Survival Guide*. New York: Three Rivers, 2003.
Brown, Hannah, and Ann Kelly. "Material Proximities and Hotspots: Towards an Anthropology of Viral Hemorrhagic Fevers." *Medical Anthropology Quarterly* 28, no. 2 (June 2014): 280–303. http://doi.org/10.1111/maq.12092.
Brydes, Linda. *Below the Magic Mountain: A Social History of Tuberculosis in Twentieth-Century Britain*. Oxford: Clarendon, 1988.
Burnet, Frank M. *Natural History of Infectious Diseases*. Cambridge: Cambridge University Press, 1972.
Butler, Declan. "Politically Correct Names Given to Flu Viruses." *Nature* 452, no. 7190 (April 2008): 923. http://doi.org/10.1038/452923a.
Butler, Declan. "Q&A: Siti Fadilah Supari." *Nature* 450, no. 1137 (December 19, 2007).
Butt, Zoe. "Voracious Embrace." Review, LenaBui.com. http://www.lenabui.com/voracious-embrace/.
Cabestan, Jean-Pierre, and Benoît Vermander. *La Chine en quête de ses frontières: La confrontation Chine-Taiwan*. Paris: Presses de Sciences Po, 2005.
Caduff, Carlo. "Anticipations of Biosecurity." In *Biosecurity Interventions: Global Health and Security in Question*, edited by Andrew Lakoff and Stephen J. Collier, 257–77. New York: SSRC–Columbia University Press, 2008.
Caduff, Carlo. *The Pandemic Perhaps: Dramatic Events in a Public Culture of Danger*. Oakland: University of California Press, 2015.
Caduff, Carlo. "Pandemic Prophecy: or How to Have Faith in Reason." *Current Anthropology* 55, no. 3 (June 2014): 296–315. https://doi.org/10.1086/676124.
Caduff, Carlo. "The Semiotics of Security: Infectious Disease Research and the Biopolitics of Informational Bodies in the United States." *Cultural Anthropology* 27, no. 2 (May 2012): 333–57. https://doi.org/10.1111/j.1548-1360.2012.01146.
Calvert, Jane. "Systems Biology, Big Science and Grand Challenges." *BioSocieties* 8, no. 4 (December 2013): 466–79.
Carrithers, Michael, Matei Candea, Karen Sykes, Martin Holbraad, and Soumhya Venkatesan. "Ontology Is Just Another Word for Culture:

Motion Tabled at the 2008 Meeting of the Group for Debates in Anthropological Theory, University of Manchester." *Critique of Anthropology* 30, no. 2 (June 2010): 152–200. https://doi.org/10.1177/0308275X09364070.

Carroll, John M. *A Concise History of Hong Kong*. Hong Kong: Hong Kong University Press, 2007.

Carter, K. Codell. *The Rise of Causal Concepts of Disease: Case Histories.* Aldershot, UK: Ashgate, 2003.

Catley, Andrew, Robin Alders, and James Wood. "Participatory Epidemiology: Approaches, Methods, Experiences." *Veterinary Journal* 191, no. 2 (February 2012): 151–60. https://doi.org/10.1016/j.tvjl.2011.03.010.

Chamayou, Grégoire. *Manhunts: A Philosophical History*. Translated by Steven Rendall. Princeton, NJ: Princeton University Press, 2012.

Charvolin, Florian, André Micoud, and Lyse Nyhart, eds. *Des sciences citoyennes? La question de l'amateur dans les sciences naturalistes*. La Tour d'Aigues: Éditions de l'Aube, 2007.

Chen, Tzung-Wen. "Global Technology and Local Society: Developing a Taiwanese and Korean Bioeconomy through the Vaccine Industry." *East Asian Science, Technology and Society* 9, no. 2 (2015): 167–86. https://doi.org/10.1215/18752160-2876770.

Cheung, C. Y., L. L. M. Poon, A. S. Lau, W. Luk, Y. L. Lau, K. F. Shortridge, S. Gordon, Y. Guan, and J. S. M. Peiris. "Induction of Proinflammatory Cytokines in Human Macrophages by Influenza a (h5n1) Viruses: A Mechanism for the Unusual Severity of Human Disease?" *Lancet* 360, no. 9348 (December 2002): 1831–37. https://doi.org/10.1016/S0140-6736(02)11772-7.

Ching, Leo. *Becoming Japanese: Colonial Taiwan and the Politics of Identity Formation*. Berkeley: University of California Press, 2001.

Chiva, Isac. "Qu'est-ce qu'un musée des arts et traditions populaires? Entretien avec Claude Lévi-Strauss." *Le Débat* 3, no. 70 (1992): 156–63.

Chou, Cynthia. "Agriculture and the End of Farming in Singapore." In *Nature Contained: Environmental Histories of Singapore*, edited by Timothy B. Barnard, 216–40. Singapore: NUS Press, 2014.

Choy, Timothy. *Ecologies of Comparison: An Ethnography of Endangerment in Hong Kong*. Durham, NC: Duke University Press, 2011.

Clifford, James. *The Predicament of Culture: Twentieth-Century Ethnography, Literature, and Art.* Cambridge, MA: Harvard University Press, 1988.

Coggins, Chris. *The Tiger and the Pangolin: Nature, Culture, and Conservation in China*. Honolulu: University of Hawai'i Press, 2002.

Colborn, Theo, Dianne Dumanoski, and John Peterson Myers. *Our Stolen Future: Are We Threatening Our Fertility, Intelligence, and Survival? A Scientific Detective Story*. New York: Dutton, 1996.

Collier, Stephen J., Andrew Lakoff, and Paul Rabinow. "Biosecurity:

Towards an Anthropology of the Contemporary." *Anthropology Today* 20, no. 5 (October 2004): 3–7. https://doi.org/10.1111/j.0268-540X.2004.00292.x.

Cooper, Melinda. "Pre-empting Emergence. The Biological Turn in the War on Terror." *Theory, Culture and Society* 23, no. 4 (July 2006): 113–35. https://doi.org/10.1177/0263276406065121.

Creager, Angela N. H. *The Life of a Virus: Tobacco Mosaic Virus as an Experimental Model, 1930–1965*. Chicago: University of Chicago Press, 2002.

Croddy, Eric. "China's Role in the Chemical and Biological Disarmament Regimes." *Nonproliferation Review* 9, no. 3 (2002): 16–47. https://doi.org/10.1080/10736700208436872.

Daston, Lorraine, and Peter Galison. *Objectivity*. New York: Zone, 2007.

Davis, Mike. *The Monster at Our Door: The Global Threat of Avian Flu*. New York: Henry Holt, 2006.

Davis, Tracy. *Stages of Emergency: Cold War Nuclear Civil Defense*. Durham, NC: Duke University Press, 2007.

de Kruif, Paul. *Microbe Hunters*. New York: Harcourt-Brace, 1926.

Delaporte, François. "Contagion et infection." In *Dictionnaire de la pensée médicale*, edited by Dominique Lecourt, 283–87. Paris: PUF, 2004.

Delgado, Christopher L., Mark W. Rosegrant, Henning Steinfeld, Simeon Ehui, and Claude Courbois. *The Coming Livestock Revolution*. New York: United Nations, 2000.

Derrida, Jacques. "The Animal That Therefore I Am (More to Follow)." Translated and edited by David Wills. *Critical Inquiry* 28, no. 2 (2002): 369–418.

Derrida, Jacques. "Autoimmunity: Real and Symbolic Suicide." In *Philosophy in a Time of Terror: Dialogues with Jürgen Habermas and Jacques Derrida*, edited by Giovanna Borradori. Chicago: University of Chicago Press, 2003.

Descola, Philippe. "Les avatars du principe de causalité." In *Les idées de l'anthropologie*, edited by Philippe Descola, Gérard Lenclud, Carlo Severi, and Anne-Christine Taylor. Paris: Armand Colin, 1988.

Descola, Philippe. *Beyond Nature and Culture*. Translated by Janet Lloyd. Chicago: University of Chicago Press, 2013.

Descola, Philippe. *The Spears of Twilight: Life and Death in the Amazon Jungle*. Translated by Janet Lloyd. New York: New Press, 1996.

Despret, Vinciane. *Naissance d'une théorie éthologique: La danse du cratérope écaillé*. Le Plessis-Robinson: Synthélabo, 1996.

Dhanasekaran, Vijaykrishna, Gavin J. D. Smith, Jing Xua Zhang, J. S. M. Peiris, Hongling Chen, and Yi Guan. "Evolutionary Insights into the Ecology of Coronaviruses." *Journal of Virology* 81, no. 15 (August 2007): 4012–20. http://doi.org/10.1128/JVI.01135-07.

Diamond, Jared. *Guns, Germs and Steel: The Fates of Human Societies*. New York: W. W. Norton, 1997.

Dikötter, Frank. *Mao's Great Famine: The History of China's Most Devastating Catastrophe*. London: Bloomsbury, 2010.
Doherty, Peter. *Sentinel Chickens: What Birds Tell Us about Our Health and the World*. Melbourne: Melbourne University Press, 2012.
Domingo, E., V. Martín, C. Perales, A. Grande-Pérez, J. García-Arriaza, and A. Arias. "Viruses as Quasispecies: Biological Implications." *Current Topics in Microbiology and Immunology* 299 (February 2006): 51–82. http://doi.org?10.1007/3-540-26397-7_3.
Drexler, Jan Felix, Victor Max Corman, and Christian Drosten. "Ecology, Evolution and Classification of Bat Coronaviruses in the Aftermath of SARS." *Antiviral Research* 101 (January 2014): 45–56. http:doi.org/10.1016/j.antiviral.2013.10.013.
Drexler, Martine. *Secret Agents: The Menace of Emerging Infections*. Washington, DC: Joseph Henry Press, 2002.
Drosten, Christian, S. Günter, W. Preiser, S. van der Werf, H. R. Brodt, S. Becker, H. Rabenau, et al. 2003. "Identification of a Novel Coronavirus in Patients with Severe Acute Respiratory Syndrome." *New England Journal of Medicine* 348, no. 20 (2003): 1967–76.
Duara, Prasenjit. "Hong Kong and the New Imperialism in East Asia 1941–1966." In *Colonialism and Chinese Localities*, edited by David Goodman and Bryna Goodman, 183–202. London: Routledge, 2009.
Dubos, René. *Man, Medicine and Environment*. London: Pall Mall Press, 1968.
Duncan, Karen. *Hunting the 1918 Flu: One Scientist's Search for a Killer Virus*. Toronto: University of Toronto Press, 2003.
Dunlap, Thomas. *In the Field, Among the Feathered: A History of Birders and Their Guides*. Oxford: Oxford University Press, 2011.
Durkheim, Émile. *Elementary Forms of Religious Life*. Translated by Joseph Ward Swain. London: Allen and Unwin, 1915.
Durkheim, Émile. *Rules of Sociological Method*. Edited by Steven Lukes. Translated by William D. Halls. New York: Free Press, 1982.
Eidson, M., N. Komar, F. Sorhage, R. Nelson, T. Talbot, F. Mostashari, R. McLean, and West Nile Virus Avian Mortality Surveillance Group. "Crow Death as a Sentinel Surveillance System for Westnile Virus in the Northern United States, 1999." *Emerging Infectious Diseases* 7, no. 4 (July 2001): 615–20. http://doi.org/10.3201/eid0704.010402.
Enemark, Christian. *Disease and Security: Natural Plagues and Biological Weapons in East Asia*. London: Routledge, 2007.
Enticott, Gareth. "Calculating Nature: The Case of Badgers, Bovine Tuberculosis and Cattle." *Journal of Rural Studies* 17, no. 2 (April 2001): 149–64. https://doi.org/10.1016/S0743-0167(00)00051-6.
Erickson, Philippe. "De l'acclimatation des concepts et des animaux, ou les tribulations d'idées américanistes en Europe." *Terrain* 28 (1997): 119–24.

Etheridge, Elizabeth. *Sentinel for Health: A History of the Centers for Disease Control.* Berkeley: University of California Press, 1992.
Evans-Pritchard, Edward E. *The Nuer: A Description of the Modes of Livelihood and Political Institutions of a Nilotic People.* Oxford: Oxford University Press, 1940.
Eyler, John M. "De Kruif's Boast: Vaccine Trials and the Construction of a Virus." *Bulletin of the History of Medicine* 80, no. 3 (February 2006): 409–38. http://doi.org/10.1353/bhm.2006.0092.
Fan, Fa-Ti. *British Naturalists in Qing China: Science, Empire, and Cultural Encounter.* Cambridge, MA: Harvard University Press, 2004.
Fassin, Didier, and Mariella Pandolfi, eds. *Contemporary States of Emergency: The Politics of Military and Humanitarian Interventions.* Cambridge, MA: MIT Press and Zone Books, 2013.
Fearnley, Lyle. "Wild Goose Chase: The Displacement of Influenza Research in the Fields of Poyang Lake, China." *Cultural Anthropology* 30, no. 1 (March 2015): 12–35. https://doi.org/10.14506/ca30.1.03.
Ferret, Carole. "Towards an Anthropology of Action: From Pastoral Techniques to Modes of Action." *Journal of Material Culture* 19, no. 3 (July 2014): 279–302. https://doi.org/10.1177/1359183514540065.
Figuié, Muriel. "Towards a Global Governance of Risks: International Health Organisations and the Surveillance of Emerging Infectious Diseases." *Journal of Risk Research* 17, no. 4 (2014): 469–83. https://doi.org/10.1080/13669877.2012.761277.
Filliozat, Jean. *Magie et médecine.* Paris: PUF, 1944.
Findlen, Paula. *Possessing Nature: Museums, Collecting, and Scientific Culture in Early Modern Italy.* Berkeley: University of California Press, 1994.
Fischer, Michael. "Biopolis: Asian Science in the Global Circuitry." *Science and Technology Study* 18, no. 3 (October 2013): 381–406. https://doi.org/10.1177/0971721813498500.
Fisher, John R. "Cattle Plagues Past and Present: The Mystery of Mad Cow Disease." *Journal of Contemporary History* 33, no. 2 (April 1998): 215–28.
Foucault, Michel. "Omnes et Singulatim: Towards a Criticism of Political Reason." *Power* 3 (January 2000): 298–25.
Foucault, Michel. *Security, Territory, Population: Lectures at the Collège de France 1977–1978.* Translated by Graham Burchell. London: Palgrave Macmillan, 2007.
Foucault, Michel. *Society Must Be Defended: Lectures at the Collège de France 1975–1976.* Translated by David Macey. New York: Picador, 2003.
Fouchier, Ron, et al. "Gain-of-Function Experiments on H7N9." *Science* 341, no. 6146 (August 2013): 612–13. http://doi.org/10.1126/science.341.6146.612.
Fox-Keller, Evelyn. *A Feeling for the Organism: The Life and Work of Barbara McClintock.* New York: W. H. Freeman, 1983.

Franklin, Sarah. *Dolly Mixtures: The Remaking of Genealogy*. Durham, NC: Duke University Press, 2007.

Frege, Gottlob. *Logical Investigations*. Translated and edited by Peter Geach. London: Blackwell, 1975.

Friedmann, Sara. *Exceptional States: Chinese Immigrants and Taiwanese Sovereignty*. Berkeley: University of California Press, 2015.

Galison, Peter. "The Future of Scenarios: State Science Fiction." In *The Subject of Rosi Braidotti: Politics and Concepts*, edited by Bolette Blaagaard and Iris van der Tuin, 38–46. London: Bloomsbury Academic, 2014.

Galison, Peter. *Image and Logic: A Material Culture of Microphysics*. Chicago: University of Chicago Press, 1997.

Gallo, Robert. *Virus Hunting: AIDS, Cancer and the Human Retrovirus: A Story of Scientific Discovery*. New York: Basic Books, 1991.

Garrett, Laurie. *The Coming Plague: Newly Emerging Diseases in a World Out of Balance*. New York: Penguin, 1995.

Gaudillière, Jean-Paul. "Rockefeller Strategies for Scientific Medicine: Molecular Machines, Viruses and Vaccines." *Studies in History and Philosophy of Science* 31, no. 3 (2000): 491–509. http://doi.org/10.1016/S1369-8486(00)00017-0.

Ghamari-Tabrizi, Sharon. *The Worlds of Herman Kahn: The Intuitive Arts of Thermonuclear War*. Cambridge, MA: Harvard University Press, 2005.

Glasse, Robert. "Cannibalisme et kuru chez les Foré de Nouvelle-Guinée." *L'Homme* 3, no. 8 (1968): 27–34.

Gorgus, Nina. *Le magicien des vitrines: Le muséologue Georges Henri Rivière*. Paris: Editions de la MSH, 2003.

Gortazar, Christian, et al. "Crossing the Interspecies Barrier: Opening the Door to Zoonotic Pathogens." *PLoS Pathogens* 10, no. 6 (June 2014). http://doi.org/10.1371/journal.ppat.1004129.

Gottweiss, Herbert. "Participation and the New Governance of Life." *Biosocieties* 3, no. 3 (September 2008): 265–86.

Gradmann, Christoph. "Robert Koch and the Invention of the Carrier State: Tropical Medicine, Veterinary Infections and Epidemiology around 1900." *Studies in History and Philosophy of Biological and Biomedical Sciences* 41 (September 2010): 232–40. http://doi.org/10.1016/j.shpsc.2010.04.012.

Gradmann, Christoph. "A Spirit of Scientific Rigour: Koch's Postulates in Twentieth-Century Medicine." *Microbes and Infection* 16, no. 11 (2014): 885–92.

Gramaglia, Christelle. "Sentinel Organisms: 'They Look out for the Environment!'" *Limn* (2013). https://limn.it/articles/sentinel-organisms-they-look-out-for-the-environment/.

Grantham, Alexander. *Via Ports: From Hong Kong to Hong Kong*. Hong Kong: Hong Kong University Press, 1965.

Greenfeld, Karl T. *China Syndrome: The True Story of the 21st Century's First Great Epidemic*. New York: HarperCollins, 2006.
Greger, Michael. *Bird Flu: A Virus of Our Own Hatching*. New York: Lantern, 2006.
Griffiths, Tom. *Hunters and Collectors: The Antiquarian Imagination in Australia*. Cambridge: Cambridge University Press, 1996.
Guan, Yi, et al. "Isolation and Characterization of Viruses Related to the SARS Coronavirus from Animals in Southern China." *Science* 302, no. 5643 (2003): 276–78.
Gusterson, Hugh. *Nuclear Rites: A Weapons Laboratory at the End of the Cold War*. Stanford, CA: Stanford University Press, 1996.
Hamayon, Roberte. *La chasse à l'âme: Esquisse d'une théorie du chamanisme sibérien*. Nanterre: Société d'ethnologie, 1990.
Hamayon, Roberte. *Why We Play: An Anthropological Study*. Translated by Damien Simon. Chicago: University of Chicago Press, 2015.
Hamblin, Jacob D. *Arming Mother Nature: The Birth of Environmental Catastrophism*. Oxford: Oxford University Press, 2013.
Handlin Smith, Joanna. "Liberating Animals in Ming-Qing China: Buddhist Inspiration and Elite Imagination." *Journal of Asian Studies* 58, no. 1 (1999): 51–84.
Hanson, Martha. *Speaking of Epidemics in Chinese Medicine: Disease and the Geographic Imagination in Late Imperial China*. London: Routledge, 2011.
Haraway, Donna. *When Species Meet*. Minneapolis: University of Minnesota Press, 2007.
Harrison, Rodney. "World Heritage Listing and the Globalization of Endangerment Sensibility." In *Endangerment, Biodiversity and Culture*, edited by Fernando Vidal and Nelia Dias, 195–217. London: Routledge, 2016.
Hathaway, Michael. *Environmental Winds: Making the Global in Southwest China*. Berkeley: University of California Press, 2013.
Heise, Ursula. "Lost Dogs, Last Birds, and Listed Species: Cultures of Extinction." *Configurations* 18, no. 1–2 (2010): 49–72.
Helmreich, Stefan. *Alien Ocean: Anthropological Voyages in Microbial Seas*. Berkeley: University of California Press, 2009.
Hinchliffe, Steve. "More than One World, More than One Health: Reconfiguring Interspecies Health." *Social Science and Medicine* 129 (2015): 28–35. http://doi.org/10.1016/j.socscimed.2014.07.007.
Hinchliffe, Steve, and Nick Bingham. "Securing Life: The Emerging Practices of Biosecurity." *Environment and Planning* 40 (2008): 1534–51.
Hinterberger, Amy, and Natalie Porter. "Genomic and Viral Sovereignty: Tethering the Materials of Global Biomedicine." *Public Culture* 27, no. 2, 76 (2015): 361–86. http://doi.org/10.1215/08992363-2841904.
Hirst, George. "The Agglutination of Red Cells by Allantoic Fluid of Chick

Embryos Infected with Influenza Virus." *Science* 94, no. 2427 (1941): 22–23. http://doi.org/10.1126/science.94.2427.22.
Hoong, Cha M. *A Defining Moment: How Singapore Beat SARS*. Singapore: Stamford Press, 2004.
Houseman, Michael. "Dissimulation and Simulation as Modes of Religious Reflexivity." *Social Anthropology* 10, no. 1 (2002): 77–89.
Houseman, Michael, and Carlo Severi. *Naven or the Other Self: A Relational Approach to Ritual Action*. Leiden: Brill, 1998.
Hsiao, Michael H. H. "Environmental Movements in Taiwan." In *Asia's Environmental Movements: Comparative Perspectives*, edited by Yok-Shiu Lee and Alvin Y. So, 32–45. Armonk, NY: M. E. Sharpe, 1999.
Huang, Michael. "Saving Pillow Mountain, Taiwan." *World Bird Watch* 22, no. 3 (2000): 10–11.
Hugh-Jones, Stephen. "Shamans, Prophets, Priests and Pastors." In *Shamanism, History and the State*, edited by N. Thomas and C. Humphrey, 32–75. Ann Arbor: University of Michigan Press, 1996.
Ingold, Tim. *Hunters, Pastoralists and Ranchers: Reindeer Economies and Their Transformations*. Cambridge: Cambridge University Press, 1980.
Ingold, Tim. *The Perception of the Environment*. New York: Routledge, 2000.
Ingold, Tim. "The Significance of Storage in Hunting Societies." *Man* 18, no. 3 (1983): 553–71.
Ingrao, Christian. *The SS Dirlewanger Brigade: The History of the Black Hunters*. New York: Skyhorse, 2011.
Investigation Group on Epidemiological Study. 2009. *Epidemiology Report of the Highly Pathogenic Avian Influenza H5N1 Outbreak in December 2008 in a Chicken Farm in Ha Tsuen, New Territories*. https://www.afcd.gov.hk/files/english/EPI_Report_Eng_v3.pdf.
James, L., N. Shindo, J. Cutter, S. Ma, and S. K. Chew. "Public Health Measures Implemented during the SARS Outbreak in Singapore, 2003." *Public Health* 120, no. 1 (2006): 20–26. https://doi.org/10.1016/j.puhe.2005.10.005.
Jones, Susan. "Mapping a Zoonotic Disease: Anglo-American Efforts to Control Bovine Tuberculosis before World War I." *Osiris* 19 (2004): 133–48.
Kahn, Herman. *Thinking about the Unthinkable*. Princeton, NJ: Princeton University Press, 1962.
Karsenti, Bruno. *Politique de l'esprit: Auguste Comte et la naissance de la science sociale*. Paris: Hermann, 2006.
Keck, Frédéric. "Assurance and Confidence in *The Two Sources of Morality and Religion:* A Sociological Interpretation of the Distinction between Static Religion and Dynamic Religion." In *Bergson, Politics, and Religion*, edited by A. Lefebvre and M. White, 265–80. Durham, NC: Duke University Press, 2012.

Keck, Frédéric. "Bergson dans la société du risque." In *Lectures de Bergson*, edited by C. Riquier and F. Worms, 164–84. Paris: PUF, 2011.
Keck, Frédéric. "Bird Flu: Are Viruses Still in the Air?" *Conversation*, 2018. https://theconversation.com/bird-flu-are-viruses-still-in-the-air-99604.
Keck, Frédéric. "'Ce virus est potentiellement pandémique': Les énoncés divinatoires des experts de la grippe aviaire." *Anthropologie et Société* 42, no. 2–3 (2018): 271–89.
Keck, Frédéric. *Claude Lévi-Strauss: Une introduction*. Paris: La Découverte-Pocket, 2011.
Keck, Frédéric. "The Contaminated Milk Scandal." *China Perspectives* 1 (2009): 88–93.
Keck, Frédéric. "Des virus émergents aux bactéries résistantes: Une crise sanitaire et ses effets." *Médecine/Sciences* 28 (2012): 534–37.
Keck, Frédéric. "Feeding Sentinels: Logics of Care and Biosecurity in Farms and Labs." *BioSocieties* 10, no. 2 (2015): 162–76.
Keck, Frédéric. "Lévi-Strauss et l'Asie: L'anthropologie structurale 'out of America.'" *EchoGéo* 7 (2008). http://journals.openedition.org/echogeo/9593; http://doi.org/10.4000/echogeo.9593.
Keck, Frédéric. "Live Poultry Markets and Avian Flu in Hong Kong." In *Food: Ethnographic Encounters*, edited by Leo Coleman, 49–58. London: Berg, 2011.
Keck, Frédéric. *Lucien Lévy-Bruhl, entre philosophie et anthropologie: Contradiction et participation*. Paris: Editions du CNRS, 2008.
Keck, Frédéric. *Un monde grippé.* Paris: Flammarion, 2010.
Keck, Frédéric, and Andrew Lakoff. "Sentinel Devices." *Limn* 3 (2013). https://limn.it/articles/preface-sentinel-devices-2/.
Keck, Frédéric, and Christos Lynteris. "Zoonosis: Prospects and Challenges for Medical Anthropology." *Medicine, Anthropology, Theory* 5, no. 3 (2018): 1–14. https://doi.org/10.17157/mat.5.3.372; http://www.medanthrotheory.org/read/10867/zoonosis.
Keck, Frédéric, Ursula Regehr, and Skaia Walentowicz. "Anthropologie: Le tournant ontologique en action." *Tsantsa* 20 (2015): 34–41.
Kelly, John D. "Introduction: The Ontological Turn in French Philosophical Anthropology." *Hau* 4 (2014): 259–69.
Kikuchi, Yuko, ed. *Refracted Modernity: Visual Culture and Identity in Colonial Taiwan*. Honolulu: University of Hawai'i Press, 2007.
Kilbourne, Edwin. "Influenza Pandemics: Can We Prepare for the Unpredictable?" *Viral Immunology* 17, no. 3 (2004): 350–57. https://doi.org/10.1089/vim.2004.17.350.
Kilbourne, Edwin. "Influenza Pandemics of the 20th Century." *Emerging Infectious Diseases* 12, no. 1 (January 2006): 9–14. https://doi.org/10.3201/eid1201.051254.

Kilburn, Mike. "Railway Development Threatens Long Valley." *World Bird Watch* 22, no. 3 (2000): 8.

King, Nicholas. "Security, Disease, Commerce: Ideologies of Postcolonial Global Health." *Social Studies of Science* 32, no. 5–6 (2002): 763–89.

Kleinman, Arthur, Barry Bloom, A. Saich, Katherine Mason, and Felicity Aulino. "Avian and Pandemic Influenza: A Biosocial Approach; Introduction." *Journal of Infectious Diseases* 197 (2008): S1–S3. https//doi.org/10.1086/524992.

Kleinman, Arthur, and James Watson, eds. *SARS in China: Prelude to Pandemics*. Stanford, CA: Stanford University Press, 2006.

Kohn, Eduardo. *How Forests Think: Toward an Anthropology beyond the Human.* Berkeley: University of California Press, 2013.

Kolata, Gina. *Flu: The Story of the Great Influenza Pandemic and the Search for the Virus That Caused It.* New York: Simon and Schuster, 1999.

Kourilsky, Philippe. *Le jeu du hasard et de la complexité: La nouvelle science de l'immunologie*. Paris: Odile Jacob, 2014.

Kowal, Emma, and Joanna Radin, eds. *Cryopolitics. Frozen Life in a Melting World.* Cambridge, MA: MIT Press, 2017.

Kuiken, Thijs, et al. "Host Species Barriers to Influenza Virus Infections." *Science* 21, 312 no. 5772 (April 2006): 394–97.

Lachenal, Guillaume. "Lessons in Medical Nihilism: Virus Hunters, Neoliberalism and the AIDS Crisis in Cameroon." In *Science and the Parastate in Africa*, edited by Wenzel Geissler, 103–41. Durham, NC: Duke University Press, 2015.

Lakoff, Andrew. "The Risks of Preparedness: Mutant Bird Flu." *Public Culture* 24, no. 368 (2012): 457–64. http://doi.org/10.1215/08992363-1630636.

Lakoff, Andrew. "Two Regimes of Global Health." *Humanity: An International Journal of Human Rights, Humanitarianism and Development* 1, no. 1 (2010): 59–80.

Lakoff, Andrew. *Unprepared: Global Health in a Time of Emergency*. Oakland: University of California Press, 2017.

Landecker, Hannah. *Culturing Life: How Cells Became Technologies.* Cambridge, MA: Harvard University Press, 2007.

Landecker, Hannah. "Food as Exposure: Nutritional Epigenetics and the New Metabolism." *BioSocieties* 6, no. 2 (June 2011): 167–94. http://doi.org/10.1057/biosoc.2011.1.

Langston, Nancy. *Toxic Bodies: Hormone Disruptors and the Legacy of DES*. New Haven, CT: Yale University Press, 2010.

Latour, Bruno. *The Pasteurization of France*. Cambridge, MA: Harvard University Press, 1993.

Laurière, Christine. *Paul Rivet: Le savant et le politique*. Paris: Publications scientifiques du Muséum national d'histoire naturelle, 2008.

Laver, Graeme. "Influenza Virus Surface Glycoproteins H and N: A Personal

Account." In *Influenza*, edited by Charles W. Potter, 31–47. Amsterdam: Elsevier, 2002.
Law, John, and Michael Lynch. "Lists, Field-guides and the Organization of Seeing: Birdwatching as an Exemplary Observational Activity." *Human Studies* 11 (1988): 271–303.
Law, John, and Annemarie Mol. "Veterinary Realities: What Is Foot and Mouth Disease?" *Sociologia Ruralis* 51, no. 1 (2011): 1–16.
Lederberg, Joshua. "Infectious History." *Science* 288 (2000): 287–93.
Lee, Benjamin N. "Poultries in Taiwan." In *Resources of Livestocks and Poultries in Taiwan*, edited by Thomas Lih and Benjamin N. Lee. Taipei: Taiwan Bank, 1952.
Lee, Chang-Chun, et al. "Emergence and Evolution of Avian H5N2 Influenza Viruses in Chickens in Taiwan." *Journal of Virology* 88, no. 10 (2014): 5677–86. http://doi.org/10.1128/JVI.00139-14.
Leiris, Michel. "La possession et ses aspects théâtraux chez les Ethiopiens du Gondar." In *Miroir de l'Afrique.* Paris: Gallimard, 1996.
Lemov, Rebecca. "Anthropological Data in Danger." In *Endangerment, Biodiversity and Culture*, edited by Fernando Vidal and Nelia Dias, 87–111. London: Routledge, 2015.
Lentzos, Filippa, and Nikolas Rose. "Governing Insecurity: Contingency Planning, Protection, Resilience." *Economy and Society* 38, no. 2 (May 2009): 230–54. https://doi.org/10.1080/03085140902786611.
Lépinay, Vincent. *Codes of Finance: Engineering Derivatives in a Global Bank.* Princeton, NJ: Princeton University Press, 2011.
Le Roy, Charles-Georges. "Lettre sur les animaux." In *Studies on Voltaire and the Eighteenth Century*, edited by Elizabeth Anderson, 316. Oxford: The Voltaire Foundation at the Taylor Institution, Oxford University, 1994.
Leung, Angela K. C. "The Evolution of the Idea of *Chuanran* (Contagion) in Imperial China." In *Health and Hygiene in Chinese East Asia: Policies and Publics in the Long Twentieth Century*, edited by Angela K. C. Leung and Christine Furth, 25–50. Durham, NC: Duke University Press, 2010.
Leung, Gabriel, and John Bacon-Shone. *Hong Kong's Health System: Reflections, Perspectives and Visions.* Hong Kong: Hong Kong University Press, 2006.
Leung, Ping-Chung. "Efficacy of Chinese Medicine for SARS." In *Bird Flu: A Rising Pandemic in Asia and Beyond*, edited by Paul Tambyah and Ping-Chung Leung, 147–66. Singapore: World Scientific, 2006.
Lévi-Strauss, Claude. "La crise moderne de l'anthropologie." *Courrier de l'UNESCO* 11 (1961): 12–18.
Lévi-Strauss, Claude. "The Lessons of Linguistics." In *The View from Afar*, translated by Joachim Neugroschel and Phoebe Hoss. Chicago: University of Chicago Press, 1985.
Lévi-Strauss, Claude. *Savage Mind.* Translated by Julian Pitt-Rivers. London: Weidenfeld and Nicolson, 1966.

Lévi-Strauss, Claude. *Les structures élémentaires de la parenté*. Paris: Mouton, 1967.
Lévi-Strauss, Claude. *Paroles données*. Paris: Plon, 1984.
Lévi-Strauss, Claude. *Totemism*. Translated by Rodney Needham. London: Merlin, 1964.
Lévi-Strauss, Claude. *Tristes tropiques*. Paris: Plon, 1955.
Lévi-Strauss, Claude. *We Are All Cannibals: And Other Essays*. Translated by Jane Marie Todd. New York: Columbia University Press, 2016.
Lévy-Bruhl, Lucien. *Primitive Mentality*. Translated by Lilian A. Clare. London: Allen and Unwin, 1923.
Lewis, Daniel. *The Feathery Tribe: Robert Ridgway and the Modern Study of Birds*. New Haven, CT: Yale University Press, 2012.
Lindenbaum, Shirley. *Kuru Sorcery: Disease and Danger in the New Guinea Highlands*. Palo Alto, CA: Mayfield, 1979.
Linfa, Wang, and Christopher Cowled. *Bats and Viruses: A New Frontier of Emerging Infectious Diseases*. New York: Wiley, 2015.
Lipsitch, Marc, and Alison P. Galvani. "Ethical Alternatives to Experiments with Novel Potential Pandemic Pathogens." *PLoS Medicine* 11, no. 5 (2014). http://doi.org/10.1371/journal.pmed.1001646.
Liu, Tik-Sang. "Custom, Taste and Science: Raising Chickens in the Pearl River Delta, South China." *Anthropology and Medicine* 15, no. 1 (2008): 7–18.
Loh, Christine. *Underground Front: The Chinese Communist Party in Hong Kong*. Hong Kong: Hong Kong University Press, 2010.
Lowe, Celia. "Viral Clouds: Becoming H5N1 in Indonesia." *Cultural Anthropology* 4 (2010): 625–49.
Lukes, Steven. *Émile Durkheim: His Life and Works*. Stanford, CA: Stanford University Press, 1995.
Lynteris, Christos. *The Ethnographic Plague: Configuring Disease on the Chinese-Russian Frontier*. London: Palgrave Macmillan, 2016.
Lynteris, Christos. "Skilled Natives, Inept Coolies: Marmot Hunting and the Great Manchurian Pneumonic Plague (1910–1911)." *History and Anthropology* 24, no. 3 (2013): 303–21.
Lynteris, Christos. "Zoonotic Diagrams: Mastering and Unsettling Human-Animal Relations." *Journal of the Royal Anthropological Institute* 23, no. 3 (2017): 463–85.
Mackenzie, Adrian. "Bringing Sequences to Life: How Bioinformatics Corporealizes Sequence Data." *New Genetics and Society* 22, no. 3 (2003): 315–32. https://doi.org/10.1080/1463677032000147180.
MacKenzie, John. *The Empire of Nature: Hunting, Conservation and British Imperialism*. Manchester, UK: Manchester University Press, 1988.
MacPhail, Theresa. *Viral Network: A Pathography of the H1N1 Influenza Pandemic*. Ithaca, NY: Cornell University Press, 2014.

Malraux, André. *Le Musée imaginaire*. Paris: Gallimard, 1947.
Manceron, Vanessa. "Recording and Monitoring: Between Two Forms of Surveillance." *Limn* 3 (2013). https://limn.it/articles/recording-and-monitoring-between-two-forms-of-surveillance/.
Manceron, Vanessa. "What Is It like to Be a Bird? Imagination zoologique et proximité à distance chez les amateurs d'oiseaux en Angleterre." In *Bêtes à pensées: Visions des mondes animaux*, edited by Michèle Cros, Julien Bondaz, and Frédéric Laugrand. Paris: Éditions des Archives contemporaines, 2015.
Manson, Elisabeth. *Infectious Change: Reinventing Chinese Public Health after an Epidemic*. Stanford, CA: Stanford University Press, 2016.
Mantovani, Alberto, Raffaella Bonecchi, and Massimo Locati. "Tuning Inflammation and Immunity by Chemokine Sequestration: Decoys and More." *Nature Reviews Immunology* 6, no. 12 (2006): 907–18. http://doi.org/10.1038/nri1964.
Mantovani, Alberto, Massimo Locati, Annunciata Vecchi, Silvano Sozzani, and Paola Allavena. "Decoy Receptors: A Strategy to Regulate Inflammatory Cytokines and Chemokines." *Trends in Immunology* 22, no. 6 (2001): 328–36. http://doi.org/10.1016/S1471-4906(01)01941-X.
Marcus, George, and Fred Myers. *The Traffic in Culture: Refiguring Art and Anthropology*. Berkeley: University of California Press, 1995.
Maris, Virginie, and Arnaud Béchet. "From Adaptive Management to Adjustive Management: A Pragmatic Account of Biodiversity Values." *Conservation Biology* 24 (August 2010): 966–73. http://doi.org/10.1111/j.1523-1739.2009.01437.x.
Martin, Emily. *Flexible Bodies: Tracking Immunity in American Culture from the Days of Polio to the Age of AIDS*. Boston: Beacon Press, 1994.
Masashi, Y., and K. Nagahisa. "In Memoriam: Elliott McClure 1910–1998." *Auk* 116, no. 4 (1999): 1125–26.
Masco, Joseph. *The Nuclear Borderlands: The Manhattan Project in Post-Cold War New Mexico*. Princeton, N.J: Princeton University Press, 2006.
McClure, Elliott. *Migration and Survival of the Birds of Asia*. Bangkok: White Lotus Press, 1974.
McCluskey, Brian, Brandy Burgess, James Glover, Hailu Kinde, and Sharon Hietala. "Use of Sentinel Chickens to Evaluate the Effectiveness of Cleaning and Disinfection Procedures in Non-Commercial Poultry Operations Infected with Exotic Newcastle Disease Virus." *Journal of Veterinary Diagnostic Investigations* 18 (May 2006): 296–99. https://doi.org/10.1177/104063870601800313.
McCormick, Joseph, and Susan Fischer Hoch. *The Virus Hunters: Dispatchers from the Frontline*. London: Bloomsbury, 1997.
McCullough, Joseph. *Zombies: A Hunter's Guide*. Oxford: Osprey, 2013.

Mehner, Maximilian. "Zombie-Survival als Zeit-Phänomen." Master's thesis, University of Marburg.

Mendelsohn, Andrew J. "'Like All That Lives': Biology, Medicine and Bacteria in the Age of Pasteur and Koch." *History and Philosophy of the Life Sciences* 24, no. 1 (2002): 3–36.

Miller, John, and Kirsten Miller. *Hong Kong: Chronicles Abroad*. San Francisco: Chronicle, 1994.

Moore, Norman W. "Indicator Species." *Nature in Focus* 14 (1973): 3–6.

Moss, Stephen. *A Bird in the Bush: A Social History of Birdwatching*. London: Aurum Press, 2004.

Moulin, Anne-Marie, ed. *L'aventure de la vaccination*. Paris: Fayard, 1996.

Moulin, Anne-Marie. *Le dernier langage de la médecine: Histoire de l'immunologie, de Pasteur au SIDA*. Paris: Presses Universitaires de France, 1991.

Moulin, Anne-Marie. "The Network of the Overseas Pasteur Institutes: Sciences and Empires." In *Sciences and Empires*, edited by Patrick Petitjean, Catherine Jami, and Anne-Marie Moulin, 307–22. Dordrecht: Kluwer Academic, 1992.

Moulin, Anne-Marie. "Preface." In *Un ethnologue chez les chasseurs de virus: Enquête en Guyane Française*, by Christophe Perrey. Paris: L'Harmattan, 2012.

Munster, V. J., et al. "Practical Considerations for High-Throughput Influenza A Virus Surveillance Studies of Wild Birds by Use of Molecular Diagnostic Tests." *Journal of Clinical Microbiology* 47, no. 3 (March 2009): 666–73. http://doi.org/ 10.1128/JCM.01625-08.

Nading, Alex. "Humans, Animals, and Health: From Ecology to Entanglement." *Environment and Society: Advances in Research* 40, no. 1 (2013): 60–78.

Napier, David. *The Age of Immunology: Conceiving a Future in an Alienating World*. Chicago: University of Chicago Press, 2003.

Narat, Victor, Lys Alcayna-Stevens, Stephanie Rupp, and Tamara Giles-Vernick. "Rethinking Human-Nonhuman Primate Contact and Pathogenic Disease Spillover." *Ecohealth* 14, no. 4 (December 2017): 840–50. http://doi.org/10.1007/s10393-017-1283-4.

Neustadt, Richard, and Harvey Feinberg. *The Epidemic That Never Was: Policy Making and the Swine Flu Scare*. New York: Vintage, 1983.

Nguyen, Vinh-Kim. *The Republic of Therapy: Triage and Sovereignty in West Africa's Time of AIDS*. Durham, NC: Duke University Press, 2010.

Ong, Aihwa, ed. *Asian Biotech: Ethics and Communities of Fate*. Durham, NC: Duke University Press, 2010.

Ong, Aihwa. *Fungible Life: Experiment in the Asian City of Life*. Durham, NC: Duke University Press, 2016.

Osterhaus, Albert. "Catastrophes after Crossing Species Barriers." *Philosophical Transactions of the Royal Society of London* 356 (2001): 791–93.

Osterholm, Michael. "Preparing for the Next Pandemic." *Foreign Affairs* 84, no. 4 (2005): 24–37.

Palese, Peter. "Don't Censor Life-Saving Science." *Nature* 481, no. 115 (January 2012). http://doi.org.10.1038/481115a.

Peckham, Robert. "Economies of Contagion: Financial Crisis and Pandemic." *Economy and Society* 42, no. 2 (2013): 226–48.

Peckham, Robert. *Epidemics in Modern Asia*. Cambridge: Cambridge University Press, 2016.

Peckham, Robert. "Game of Empires: Hunting in Treaty-Port China." In *Eco-Cultural Networks and the British Empire*, edited by James Beattie, Edward Melillo, and Emily O'Gorman, 202–32. New York: Bloomsbury, 2014.

Peckham, Robert. "Matshed Laboratory: Colonies, Cultures, and Bacteriology." In *Imperial Contagions: Medicine, Hygiene, and Cultures of Planning in Asia*, edited by Robert Peckham, 123–47. Hong Kong: Hong Kong University Press, 2013.

Pedersen, Morton. *Not Quite Shamans: Spirit Worlds and Political Lives in Northwest Mongolia*. Ithaca, NY: Cornell University Press, 2011.

Peiris, J. S. Malik. "Japanese Encephalitis in Sri Lanka: The Study of an Epidemic; Vector Incrimination, Porcine Infection and Human Disease." *Transactions of the Royal Society of Tropical Medicine and Hygiene* 86, no. 3 (1992): 307–13.

Peiris, J. S. Malik, S. T. Lai, L. L. Poon, Y. Guan, L. Y. Yam, W. Lim, J. M. Nicholls, W. K. Yee, et al. "Coronavirus as a Possible Cause of Severe Acute Respiratory Syndrome." *Lancet* 361, no. 9366 (April 2003): 1319–25.

Peiris, J. S. Malik, Connie Y. Leung, and John M. Nicholls. "Innate Immune Responses to Influenza A H5N1: Friend or Foe?" *Trends in Immunology* 12 (December 2009): 574–84. http://doi.org/10.1016/j.it.2009.09.004.

Peiris, J. S. Malik, Leo L. Poon, John M. Nicholls, and Yi Guan. "The Role of Influenza Virus Gene Constellation and Viral Morphology on Cytokine Induction, Pathogenesis and Viral Virulence." *Hong Kong Medical Journal* 15, no. 3 (2009): 21–23.

Peiris, J. S. Malik, and James S. Porterfield. "Antibody-Mediated Enhancement of Flavivirus Replication in Macrophage-like Cell Lines." *Nature* 282, no. 5738 (1979): 509–11.

Petryna, Adriana. "The Origins of Extinction." *Limn* 3 (2013). https://limn.it/articles/the-origins-of-extinction/.

Pickering, William F. S. *Durkheim's Sociology of Religion: Themes and Theories*. Boston: Routledge and Kegan Paul, 1984.

Porcher, Jocelyne. *Eleveurs et animaux, réinventer le lien*. Paris: Presses Universitaires de France, 2002.

Porter, Natalie. "Bird Flu Biopower: Strategies for Multispecies Coexistence in Viêt Nam." *American Ethnologist* 40, no. 1 (2013): 132–48.
Porter, Natalie. "Ferreting Things Out: Biosecurity, Pandemic Flu and the Transformation of Experimental Systems." *Biosocieties* 11 (2016): 22–45.
Powell, D. G., K. L. Watkins, P. H. Li, and K. Shortridge. "Outbreak of Equine Influenza among Horses in Hong Kong during 1992." *Veterinary Record* 136, no. 21 (May 1995): 531–36. http://doi.org/10.1136/vr.136.21.531.
Pradeu, Thomas. *The Limits of the Self: Immunology and Biological Identity*. New York: Oxford University Press, 2012.
Price, Sally. *Paris Primitive: Jacques Chirac's Museum on the Quai Branly*. Chicago: University of Chicago Press, 2007.
Quammen, David. *Spillover: Animal Infections and the Next Human Pandemic*. New York: W. W. Norton, 2012.
Rabinow, Paul. *Anthropos Today: Reflections on Modern Equipment*. Princeton, NJ: Princeton University Press, 2003.
Rabinow, Paul. "Artificiality and Enlightenment: From Sociobiology to Biosociality." In *Essays on the Anthropology of Reason*, 91–111. Princeton, NJ: Princeton University Press, 1996.
Rabinow, Paul. *French DNA: Trouble in Purgatory*. Chicago: University of Chicago Press, 1999.
Rabinow, Paul. *Making PCR: A Story of Biotechnology*. Chicago: University of Chicago Press, 1996.
Rabinow, Paul. Preface to *Object Atlas: Fieldwork in the Museum*. Edited by Clementine Deliss. Kerber: Bielefeld, 2012
Rabinowitz, Peter, Zimra Gordon, Daniel Chudnov, Matthew Wilcox, Lynda Odofin, Ann Liu, and Joshua Dein. "Animals as Sentinels of Bioterrorism Agents." *Emerging Infectious Diseases* 12, no. 4 (2006): 647–52. http://doi.org/10.3201/eid1204.051120.
Radin, Joanna. *Life on Ice: A History of New Uses for Cold Blood*. Chicago: University of Chicago Press, 2017.
Rajan, Kaushik S. *Biocapital: The Constitution of Postgenomic Life*. Durham, NC: Duke University Press, 2006.
Rawls, Ann. *Epistemology and Practice: Durkheim's "The Elementary Forms of Religious Life."* Cambridge: Cambridge University Press, 2005.
Redfield, Peter. *Life in Crisis: The Ethical Journey of Doctors without Borders*. Berkeley: University of California Press, 2013.
Revet, Sandrine. "'A Small World': Ethnography of a Natural Disaster Simulation in Lima, Peru." *Social Anthropology/Anthropologie Sociale* 21, no. 1 (2013): 1–16. doi.org/10.1111/1469-8676.12002.
Robertson Smith, William. *The Religion of the Semites*. New York: Macmillan, [1889] 1927.
Robin, Libby. *The Flight of the Emu: A Hundred Years of Australian Ornithology 1901–2001*. Melbourne: Melbourne University Press, 2001.

Roitman, Janet. "The Garrison-Entrepôt: A Mode of Governing in the Chad Basin." In *Global Assemblages: Technology, Politics, and Ethics as Anthropological Problems*, edited by Aihwa Ong and Stephen J. Collier, 417–35. Malden, MA: Wiley-Blackwell, 2004.

Rollet, Vincent. "Dimensions identitaire, sécuritaire et sociétale de la politique étrangère de Taiwan dans le domaine de la lutte contre les maladies infectieuses (2000–2008)." PhD diss., Institut d'études politiques de Paris, Sciences Po, 2010.

Rosenkrantz, Barbara G. "The Trouble with Bovine Tuberculosis." *Bulletin of the History of Medicine* 59, no. 2 (summer 1985): 155–75.

Roustan, Mélanie. "Des clefs des réserves aux mots-clefs des bases de données: Mutations du rapport aux objets pour les conservateurs du MAAO au musée du quai Branly." In *Le tournant patrimonial: Mutations contemporaines des métiers du patrimoine*, edited by Christian Hottin and Claudie Voisenat, 117–39. Paris: Editions de la MSH, 2016.

Roy, Denny. *Taiwan: A Political History*. Ithaca, NY: Cornell University Press, 2003.

Russell, Colin A., Judith M. Fonville, André E. X. Brown, David F. Burke, David L. Smith, Sarah L. James, and Sander Herfst. "The Potential for Respiratory Droplet Transmissible A/H5N1 Influenza Virus to Evolve in a Mammalian Host." *Science* 336, no. 6088 (June 2012): 1541–47.

Sahlins, Marshall. *Stone Age Economics*. London: Tavistock, 1972.

Salomon, Rachelle, Erich Hoffmann, and Robert G. Webster. "Inhibition of the Cytokine Response Does Not Protect against Lethal H5N1 Influenza Infection." *PNAS* 104, no. 30 (July 2007): 12479–81. http://doi.org/10.1073/pnas.0705289104.

Samimian-Darash, Limor. "Practicing Uncertainty: Scenario-Based Preparedness Exercises in Israel." *Cultural Anthropology* 3, no. 3 (2016): 359–86.

Schaeffer, Jean-Marie. *L'expérience esthétique*. Paris: Gallimard, 2015.

Schaffner, Spencer. *Binocular Vision: The Politics of Representation in Birdwatching Field Guides*. Amherst: University of Massachusetts Press, 2011.

Schüll, Natasha D. "The Gaming of Chance: Online Poker Software and the Potentialization of Uncertainty." In *Modes of Uncertainty: Anthropological Cases*, edited by Limor Samimian-Darash and Paul Rabinow, 46–66. Chicago: University of Chicago Press, 2015.

Schwartz, Maxime. *How the Cows Turned Mad: Unlocking the Mysteries of Mad Cow Disease*. Translated by Etienne Schneider. Berkeley: University of California Press, 2003.

Scoones, Ian, ed. *Avian Influenza: Science, Policy and Politics*. New York: Earthscan, 2010.

Severinghaus, Sheldon, ed. *The Avifauna of Taiwan*. Taipei: Taiwan's Council of Agriculture's Forestry Bureau, 2010.

Severinghaus, Lucia Liu, Stephen K. W. Kang, and Peter S. Alexander. *A Guide to the Birds of Taiwan*. Taipei: China Post, 1970.

Sexton, Christopher. *The Life of Sir Macfarlane Burnett*. Oxford: Oxford University Press, 1991.

Shapiro, Judith. *Mao's War against Nature: Politics and the Environment in Revolutionary China*. Cambridge: Cambridge University Press, 2001.

Shi Zhengli and Hu Zhihong. "A Review of Studies on Animal Reservoirs of the SARS Coronavirus." *Virus Research* 133 (2008): 74–87. http://doi.org/10.1016/j.virusres.2007.03.012.

Shortridge, Kennedy F. "Avian Influenza Viruses in Hong Kong: Zoonotic Considerations." *Wageningen UR Frontis* 8 (2005): 9–18.

Shortridge, Kennedy F., Malik Peiris, and Yi Guan. "The Next Influenza Pandemic: Lessons from Hong Kong." *Journal of Applied Microbiology* 94 (2003): 70–79.

Shortridge, Kennedy F., and Charles H. Stuart-Harris. "An Influenza Epicentre?" *Lancet* 2 (1982): 812–13.

Silbergeld, Ellen K. *Chickenizing Farms and Food: How Industrial Meat Production Endangers Workers, Animals and Consumers*. Baltimore: Johns Hopkins University Press, 2016.

Simon, Scott. *Sadyaq balae!: L'autochtonie formosane dans tous ses états*. Québec: Presses de l'Université Laval, 2012.

Sims, L. D., T. M. Ellis, K. K. Liu, K. Dyrting, H. Wong, M. Peiris, Y. Guan, and K. F. Shortridge. "Avian Influenza Outbreaks in Hong Kong, 1997–2002." *Avian Disease* 47, no. 3 (2003): 832–38.

Singer, Peter. *Animal Liberation: A New Ethics for Our Treatment of Animals*. New York: Harper and Row, 1975.

Sipress, Alan. *The Fatal Strain: On the Trail of the Avian Flu and the Coming Pandemic*. New York: Viking, 2009.

Sismondo, Sergio. "Models, Simulations and Their Objects." *Science in Context* 12, no. 2 (summer 1999): 247–60. https://doi.org/10.1017/S0269889700003409.

Smith, Gavin J. D., X. H. Fan, J. Wang, K. S. Li, K. Qin, J. X. Zhang, D. Vijaykrishna, et al. "Emergence and Predominance of an H5N1 Influenza Vdariant in China." *PNAS* 103, no. 45 (2006): 16936–41. https://doi.org/10.1073/pnas.0608157103.

Smith, Gavin J. D., Justin Bahl, Vijaykrishna Dhanasekaran, Jinxia Zhang, Leo L. M. Poon, Honglin Chen, Robert G. Webster, J. S. Malik Peiris, and Yi Guan. "Dating the Emergence of Pandemic Influenza Viruses." *PNAS* 106, no. 28 (May 2009): 11709–12. https://doi.org/10.1073/pnas.0904991106.

Smith, Gavin J. D., Vijaykrishna Dhanasekaran, Justin Bahl, Samantha J. Lycett, Michael Worobey, Oliver G. Pybus, Siu Kit Ma, et al. "Origins and Evolutionary Genomics of the 2009 Swine-Origin H1N1 Influenza A Epidemic." *Nature* 459 (June 2009): 1122–25.

Sodikoff, Genese, ed. *The Anthropology of Extinction: Essays on Culture and Species Death*. Bloomington: Indiana University Press, 2012.
Specter, Madeline. “Nature’s Bioterrorist: Is There Any Way to Prevent a Deadly Avian-Flu Pandemic?” *New Yorker*, February 28, 2005, 50–61.
Spencer, Herbert. *Study of Sociology*. New York: Appleton, 1873.
Sperber, Dan. *Explaining Culture: A Naturalistic Approach*. Oxford: Blackwell, 1996.
Spinage, Charles. *Cattle Plague: A History*. New York: Kluwer, 2003.
Steinman, Ralph M., and Zanvil A. Cohn. “Identification of a Novel Cell Type in Peripheral Lymphoid Organs of Mice.” *Journal of Experimental Medicine* 137 (May 1973): 1142–62.
Stépanoff, Charles. *Chamanisme, rituel et cognition chez les Touvas (Sibérie du Sud)*. Paris: Editions FMSH, 2014.
Stépanoff, Charles. “Devouring Perspectives: On Cannibal Shamans in Siberia.” *Inner Asia* 11 (2009): 283–307.
Stirling, Andy C., and Ian Scoones. “From Risk Assessment to Knowledge Mapping: Science, Precaution and Participation in Disease Ecology.” *Ecology and Society* 14, no. 2 (2009): 14.
Stocking, George. *After Tylor: British Social Anthropology, 1888–1951*. London: Athlone, 1995.
Stoczkowski, Wiktor. *Anthropologies rédemptrices: Le monde selon Lévi-Strauss*. Paris: Hermann, 2008.
Strasser, Bruno. “The Experimenter’s Museum: GenBank, Natural History, and the Moral Economies of Biomedicine.” *Isis* 102 (2011): 60–96.
Striffler, Ben. *Chicken: The Dangerous Transformation of America’s Favorite Food.* New Haven, CT: Yale University Press, 2005.
Strivay, Lucienne. “Taxidermies: Le trouble du vivant.” *Anthropologie et Sociétés* 39, no. 1–2 (2015): 251–68.
Szonyi, Michael. *Cold War Island: Quemoy on the Front Line*. Cambridge: Cambridge University Press, 2008.
Takada, Ayato, and Yoshihiro Kawaoka. “Antibody-Dependent Enhancement of Viral Infection: Molecular Mechanisms and *in vivo* Implications.” *Reviews in Medical Virology* 13 (November 2003): 387–98.
Tambyah, Paul, and Ping-Chung Leung, eds. *Bird Flu: A Rising Pandemic in Asia and Beyond.* Singapore: World Scientific, 2006.
Tang, Shui-Yan, and Tang Ching-Ping. “Local Governance and Environmental Conservation: Gravel Politics and the Preservation of an Endangered Bird Species in Taiwan.” *Environment and Planning A* 36 (2004): 173–89.
Taubenberger, Jeffery K., Ann H. Reid, Amy E. Krafft, Karen E. Bijwaard, and Thomas G. Fanning. “Initial Genetic Characterization of the 1918 ‘Spanish’ Influenza Virus.” *Science* 275, no. 5307 (March 1997): 1793–96. doi: 10.1126/science.275.5307.1793.
Testart, Alain. “The Significance of Food Storage among Hunter-Gatherers:

Residence Patterns, Population Densities, and Social Inequalities." *Current Anthropology* 23 (1982): 523–37.
Testart, Alain. "Some Major Problems on the Social Anthropology of Hunter-Gatherers." *Current Anthropology* 29 (1988): 1–13.
Thomas, Keith. *Man and the Natural World: Changing Attitudes in England 1500–1800*. London: Allen Lane, 1983.
Tomes, Nancy. *The Gospel of Germs: Men, Women, and the Microbe in American Life*. Cambridge, MA: Harvard University Press, 1998.
Trémon, Anne-Christine. "*Yingti/Ruanti* (Hardware/Software): La création d'un centre culturel hakka à Taiwan." *Gradhiva* 16 (2012): 131–55.
Tsing, Anna L. *Friction: An Ethnography of Global Connection*. Princeton, NJ: Princeton University Press, 2005.
Tsing, Anna L. *The Mushroom at the End of the World: On the Possibility of Life in Capitalist Ruins*. Princeton, NJ: Princeton University Press, 2015.
Turkle, Sherry. *Simulation and Its Discontents*. Cambridge, MA: MIT Press, 2009.
Turnbull, Constance M. *A History of Singapore 1819–1975*. London: Oxford University Press, 1977.
United Nations System Influenza Coordination (UNSIC). *Simulation Exercises on Influenza Pandemic Responses in the Asia-Pacific Region*. 2008.
Vagneron, Frédéric. "Surveiller et s'unir? Le rôle de l'OMS dans les premières mobilisations internationales autour d'un réservoir animal de la grippe." *Revue d'anthropologie des connaissances* 9, no. 2 (2015): 139–62.
Van Dooren, Tom. *Flight Ways: Life and Loss at the Edge of Extinction*. New York: Columbia University Press, 2014.
Veblen, Thorstein. *The Theory of the Leisure Class*. New York: Viking Penguin, [1899] 1967.
Veríssimo, Diogo, Iain M. Fraser, Jim J. Groombridge, Rachel M. Bristol, and Douglas C. MacMillan. "Birds as Tourism Flagship Species: A Case Study of Tropical Islands." *Animal Conservation* 12 (2009): 549–58. https://doi.org/10.1111/j.1469-1795.2009.00282.x.
Vidal, Fernando, and Nelia Dias. "Introduction: The Endangerment Sensibility." In *Endangerment, Biodiversity and Culture*, edited by Fernando Vidal and Nelia Dias, 1–40. London: Routledge, 2016.
Viveiros de Castro, Eduardo. *Cannibal Metaphysics: For a Post-Structural Anthropology*. Translated by Peter Skafish. Minneapolis: University of Minnesota Press, 2014.
Viveiros de Castro, Eduardo. "Cosmological Deixis and Amerindian Perspectivism." *Journal of the Royal Anthropological Institute* 4 (1998): 469–88.
Viveiros de Castro, Eduardo. *From the Enemy's Point of View: Humanity and Divinity in an Amazonian Society*. Chicago: University of Chicago Press, 1992.

Wain-Hobson, Simon. "H5N1 Viral Engineering Dangers Will Not Go Away." *Nature* 495 (March 28, 2013). http://doi.org/10.1038/495411a.
Waldby, Catherine. "Stem Cells, Tissue Cultures and the Production of Biovalue." *Health* 6, no. 3 (2002): 305–23.
Wallace, Rodrick, Deborah Wallace, and Robert G. Wallace. *Farming Human Pathogens: Ecological Resilience and Evolutionary Process*. New York: Springer, 2009.
Webby, Richard, and Robert G. Webster. "Are We Ready for Pandemic Influenza?" *Science* 302, no. 5650 (November 2003): 1519–22.
Webster, Robert G. "William Graeme Laver: 1929–2008." *Biographical Memoirs of the Fellows of the Royal Society* 56 (2010): 215–36.
Wei, S.-H, J. R. Yang, H. S. Wu, M. C. Chang, J. S. Lin, C. Y. Lin, Y. L. Liu, et al. "Human Infection with Avian Influenza A H6N1 Virus: An Epidemiological Analysis." *Lancet Respiratory Medicine* (November 2013). http://doi.org/10.1016/S2213-2600(13): 70221-2.
Weiss, Robin A., and Anthony J. MacMichael. "Social and Environmental Risk Factors in the Emergence of Infectious Diseases." *Nature Medicine Supplement* 10 (December 2004): 70–76.
Weller, Robert. *Discovering Nature: Globalization and Environmental Culture in China and Taiwan*. Cambridge: Cambridge University Press, 2006.
Whitney, Kristoffer. "Domesticating Nature? Surveillance and Conservation of Migratory Shorebirds in the 'Atlantic Flyway.'" *Studies in History and Philosophy of Biological and Biomedical Sciences* 45, no. 1 (March 2014): 78–87. https://doi.org/10.1016/j.shpsc.2013.10.008.
Wilkinson, Louise. *Animals and Disease: An Introduction to the History of Comparative Medicine*. Cambridge: Cambridge University Press, 1992.
Willerslev, Rane. *Soul Hunters: Hunting, Animism, and Personhood among the Siberian Yukaghirs*. Berkeley: University of California Press, 2007.
Williams, Greer. *Virus Hunters: The Lives and Triumphs of Great Medical Pioneers*. London: Hutchinson, 1960.
Wilson, Robert M. *Seeking Refuge: Birds and Landscapes of the Pacific Flyway*. Seattle: University of Washington Press, 2010.
Wolfe, Nathan D. *The Viral Storm: The Dawn of a New Pandemic Age*. New York: St. Martin's Press, 2012.
Wolfe, Nathan D., Peter Daszak, A. Marm Kilpatrick, and Donald S. Burke. "Bushmeat Hunting, Deforestation, and Prediction of Zoonoses Emergence." *Emerging Infectious Diseases* 11, no. 12 (December 2005): 1822–27.
Wolfe, Nathan D., Claire P. Dunavan, and Jared Diamond. "Origins of Major Human Infectious Diseases." *Nature* 447 (May 2007): 279–83.
Woo, Patrick C. Y., Susanna K. P. Lau, and Kwok-Yung Yuen. "Infectious Diseases Emerging from Chinese Wetmarkets: Zoonotic Origins of Severe Respiratory Viral Infections." *Current Opinion in Infectious Diseases* 19, no. 5 (October 2006): 401–7.

Woods, Abigail. *A Manufactured Plague: The History of Foot-and-Mouth Disease in Britain*. London: Earthscan, 2004.
Worboys, Michael. *Spreading Germs: Disease Theories and Medical Practice in Britain 1865–1900*. Cambridge: Cambridge University Press, 2000.
World Health Organization (WHO). "Influenza." http://www.who.int/influenza/human_animal_interface/en/.
Wylie, Sara. "Hormone Mimics and Their Promise of Significant Otherness." *Science as Culture* 21, no. 1 (2011): 49–76.
Wylie, Sara, Kim Schultz, Deborah Thomas, Chris Kassotis, and Susan Nagel. "Inspiring Collaboration: The Legacy of Theo Colborn's Transdisciplinary Research on Fracking." *New Solutions: A Journal of Environmental and Occupational Health Policy* 26, no. 3 (2016): 360–88.
Yanni, Carla. *Nature's Museums: Victorian Science and the Architecture of Display*. London: Athlone, 1999.
Yeung, Edwin. "Poultry Farming in Hong Kong." Unpublished undergraduate essay, Department of Geography and Geology, University of Hong Kong, 1956.
Youatt, Ralph. "Counting Species: Biopower and the Global Biodiversity Census." *Environmental Values* 17 (2008): 393–417.
Yuen, Kwok-Yung. 1998. "Clinical Features and Rapid Viral Diagnosis of Human Disease Associated with Avian Influenza A H5N1 virus." *Lancet* 351, no. 9101: 467–71.
Zahavi, Amotz. "Mate Selection: A Selection for a Handicap." *Journal of Theoretical Biology* 53 (1975): 205–13.
Zahavi, Amotz, and Avishag Zahavi. *The Handicap Principle: A Missing Piece of Darwin's Puzzle*. Oxford: Oxford University Press, 1997.
Zhang, Joy, and Michael Barr. *Green Politics in China: Environmental Governance and State-Society Relations*. London: Pluto, 2013.
Zito, Angela. *Of Body and Brush: Grand Sacrifice and Text Performance in Eighteenth-Century China*. Chicago: University of Chicago Press, 1997.
Zylberman, Patrick. *Tempêtes microbiennes: Essai sur la politique de sécurité sanitaire dans le monde transatlantique*. Paris: Gallimard, 2013.

INDEX

adjuvants, 142, 171
aesthetic, 52, 56, 60, 61, 62, 106, 171, 176, 199n113
Agriculture and Veterinary Authority of Singapore, 117
Agriculture, Fisheries and Conservation Department, Hong Kong, 71, 72, 91, 123
Amazonia, 26, 39, 60
American Museum of Natural History, 55, 149
Animal Health Research Institute of Taiwan, 140
animal release (*fangsheng*), 80, 127, 161
animism, 27, 28
anthrax, 128
Anthropocene, 157
anticipation, 43, 46
Antigone project, 29–31, 35, 37, 40, 41, 178, 184n1
antivirals (Tamiflu, Relenza), 145
Argentina, 13
artifacts, 121, 128, 132, 135, 138, 168, 173, 177
Asia-Pacific Economic Cooperation, 119
Asilomar conference, 33
Australia, 3, 41–43, 49, 51, 71, 82, 88, 109, 167
Auxiliary Medical Service (Hong Kong), 122
avian influenza, bird flu (H5N1, H5N2, H7N7, H7N9), 30, 35, 36, 37, 41, 43, 59, 70, 71, 77, 79–81, 83, 86, 87, 89, 90, 98, 101–3, 109, 117, 119, 120, 138, 141–44, 160, 174, 178

bacteria, 16, 62, 120, 167
Bahl, Justin, 110–14
Banks, Joseph, 52, 53

Bargheer, Stefan, 52
Barnes, Charles, 89
Barretto, Ruy, 154
Bateson, Gregory, 134
bats, 2, 40, 41, 84, 187n56, 211n6
Bergson, Henri, 23, 24
Berlin Naturkünde Museum, 53, 56
Berndt, Ronald and Catherine, 24
Beutler, Bruce, 100
biodiversity, 45, 89, 91, 96, 153, 154, 160
bioinformatics, 50, 112, 114
Biopolis, 111
biosafety, 33, 97, 99
biosecurity, 5, 33–35, 73, 167, 170, 173, 176
biosociality, 170
bioterror, 128
biovalue, 139, 158, 160, 170
bird flu. *See* avian influenza
BirdLife International, 90, 94, 150
bird races, 124
black-faced spoonbills, 125–27, 165
botulism, 125
bovine tuberculosis, 15, 182n33
Brazil, 41
breeders, farmers, 12, 15, 21, 43, 69–78
bricolage, 170
British Museum, 52
Buddhism, 80, 161–63
Buffon, Georges-Louis Leclerc, Comte de, 52
Bùi, Lêna, 31
Burnet, Frank Macfarlane, 36, 49, 51

Caduff, Carlo, 47, 48
Calmette, Albert, 22
Cambridge University, 15, 35
Canada, 84, 103, 104
cancer, 103, 104, 198n101, 198n103
cannibalism, 24, 25, 134, 174
Carson, Rachel, 103
cattle/cows, 14, 23, 26, 72, 79, 159, 169
cattle plague, 13
causality, 5, 12, 13, 14, 17, 23, 27, 158, 167, 177
Centre for Food Safety (Hong Kong), 86
Centre for Health Protection (Hong Kong), 72, 83, 119, 121
Chan, Margaret, 79, 86
Chapman, Frank, 53, 56, 124
Chapple, John, 88
Chen, Peter, 93
Chen Shui-bian, 94
Chernobyl, 131
Cheung Sha Wan, 78, 160
Chiang Ching-Kuo, 150
Chiang Kai-shek, 93, 129, 150
Childe, Gordon, 167
China, 3, 41, 49, 57, 74–76, 80, 81, 88, 92, 119, 143, 146, 149, 158, 159
Chirac, Jacques, 59
Churchill, Winston, 132
citizen science, 124, 155
civil defense, 132, 133, 177
Clifford, James, 62
climate change, 175
Colborn, Theo, 103, 105, 106
Cold War, 49, 94, 109, 129, 132, 133, 170, 171, 174, 192n25, 196n79
collections, collectors, 4, 5, 11, 25, 28, 34, 36, 38, 40–42, 46, 49, 50, 51, 55, 58, 81, 83, 172, 174
Collège de France, 25, 60, 134, 136
colonialism and colonies, 41, 42, 80, 88–90, 92, 119, 140, 153, 174, 192n25
communication, 25, 27, 43, 72, 105, 106, 119, 134, 138, 141, 166, 176, 178, 180n7
compensation, 14
Comte, Auguste, 12, 19, 26
contagion, 1, 15, 17, 20, 115, 117, 143
Cook, James, 52–54
cowpox, 20
crocodile, 135, 160
culling, 5, 11, 78, 82, 117, 118, 120, 160, 173
Cutter, Jeffery, 115
Cuvier, Georges, 52

cynegetic practices, 6, 7, 28, 56, 166, 176, 178. *See also* hunting/hunter-gatherers
cytokine storm, 101

Darwin/Darwinism, 36, 53, 54, 105, 114, 177
data/data banks/data bases, 11, 15, 35, 42, 50, 51, 89, 91, 110, 112, 113, 117, 125, 140, 147, 148, 150, 152, 153, 154, 166, 173, 174, 178, 190n52, 212n58
David, Armand, 148
Davis, Mike, 157
Davis, Tracy, 132–33
Dayhoff, Margaret, 50, 51
decoy, 102, 114, 126, 130, 134, 135, 173
de Gaulle, Charles, 60
de Kruif, Paul, 37
Delaporte, François, 17
dengue virus, 98, 101, 102
Descola, Philippe, 27, 81, 134, 180n11, 183n48, 183n49, 205n62
Dhanasekaran, Vijaykrishna, 110–13, 147
Diamond, Jared, 38, 157, 169
Dias, Nelia, 58
Drosten, Christian, 41
drylab/wetlab, 147
dual use research, 33
Dubos, René, 157
Duke NUS Graduate Medical School, 109
Dumont, Louis, 61
Durkheim, Émile, 5, 19–24

Ebola, 2, 36, 38, 41, 98
ecology of infectious diseases, 49, 52, 177, 178
Eliade, Mircea, 39
emerging infectious diseases, 2, 22, 25, 38, 41, 83, 87, 98, 158, 166
endocrine disruptors, 103–6, 175
epidemiology, 13, 43, 47, 177, 178, 190n2, 191n14, 201n8
epizootics, 169, 181n16, 182n33
equine influenza, 82
equity, 6, 138, 166, 176
Erasmus Medical Centre (Rotterdam), 29, 40
Ethiopia, 136, 137
Evans-Pritchard, Edward, 13, 169
exchange, 140, 148, 154, 156, 159, 163, 174, 175. *See also* gifts; reciprocity
exercise, 63, 108, 116–29
experts, 15, 21, 43, 44, 72, 75, 89, 116, 132, 142, 155
extinction, 56, 57, 62, 64, 92, 109, 124, 175, 177

Fan, Fa-ti, 148, 149
Farrar, Jeremy, 31
Fédération Nationale des Chasseurs, 53
ferrets, 31, 32, 47
fiction, 106, 123, 129, 130, 137, 138
Filliozat, Jean, 136
Fisher-Hoch, Susan, 38
Findlen, Paula, 58
flagship species, 96
flocks, 13, 57, 157
flyways, 88, 89, 176, 195n55, 203n36
Foot-and-Mouth Disease, 12, 13
Foucault, Michel, 81, 173, 174
Fouchier, Ron, 29–37, 43
France, 21, 22, 41, 52, 56, 75
Francis, Thomas, 47
Franklin, Sarah, 158
Frazer, James, 15, 18
freezing, 142, 167, 171
Frege, Gottlob, 16, 18
Fry, Michael, 103

gain-of-function research, 33, 34
Gajdusek, Daniel Carleton, 24, 134
Galison, Peter, 130
Gallo, Robert, 38
Galvani, Alison, 33
game (hunting), 53, 56, 137–38

games, 31, 41, 114, 123–25, 129, 131, 132, 136, 137–38
GenBank, 50, 112, 125, 147
Germany, 41, 56
Ghana, 41
ghosts, 156, 165, 167, 169, 212n65
gifts, 6, 17, 157, 158, 164
Gillen, Francis James, 19
Gladwell, Malcolm, 79
Glasse, Robert, 25
global art, 45
global health, 1, 43, 44, 177
Goad, Walter, 50, 51
Goh Chok Tong, 116
Good, George, 54
grey-faced buzzard, 150
Griaule, Marcel, 60
Guan, Yi, 83, 85, 86, 109, 110
Guandu (Taiwan), 95
Guérin, Camille, 22
Gusterson, Hugh, 131, 132

Hamayon, Roberte, 136–38
Hardy, Rev. E. J., 149
hemagglutination, 47
Hendra virus, 41
Hirst, George, 47, 51
HIV/AIDS, 38
Hoffmann, Jules, 100
Holland, 41, 146
Hong Kong Birdwatching Society, 88–92, 96
Hong Kong Buddhist Association, 80
Hong Kong Poultry Farmers Association, 70, 74
hotspot, of insect infestation, 62
Houseman, Michael, 134, 135
Hsiao, Hsin-Huang Michael, 94
Huben (Taiwan), 95, 96, 97
Hu Jintao, 83
Hunter, William, 53
hunting/hunter-gatherers, 4–7, 23, 31, 37–39, 46, 50, 52, 53, 55–57, 64, 102, 104, 108, 133, 137, 140, 155, 156, 165, 167, 170, 173, 174

Ichida, Noritaka, 150
illusion, 12, 15
imagination, 23, 27, 28, 46, 58, 64, 83, 109, 113, 115, 125
imitation, 137
immunity and immunology, 21, 30, 31, 35, 41,64, 69, 71, 73, 97, 100–102, 157, 166, 174
indicator species, 56
Indonesia, 88, 146, 187n65, 195n60
infection, 1, 17, 99, 115
information, 33, 34, 37, 40, 42, 45, 46, 58, 60, 61, 81, 84, 90, 99–102, 106, 110, 114–16, 119, 147, 148, 152–54, 156, 158, 169–72, 175, 177, 197n87, 199n113
Ingold, Tim, 168
insects, 62, 75
intention, 33, 36, 37, 46, 102, 130, 178, 183n37, 184n12, 199n113
International Council of Museums (ICOM), 60, 177
International Union for the Conservation of Nature (IUCN), 56, 60, 125, 177
interventions, 4, 5, 14, 15, 19, 21, 27

Japan, 86, 88, 93, 98, 125, 140, 144, 149–52, 164, 180n7, 193n39, 208n21
Japanese encephalitis, 83, 88
Jardine Matheson, 140
Jenner, Edward, 20
Jiang Zemin, 84, 86
journalists (media), 12, 15, 70, 74, 116, 120, 128

Kadoorie Farm (Hong Kong), 160
Kahn, Herman, 130
Kaplan, Martin, 42, 49
Kawaoka, Yoshihiro, 32–36, 43
Kerchache, Jacques, 61
Kilbourne, Edwin, 143

Kilburn, Mike, 91, 153, 154, 161, 195n63, 196n66, 210n41
King, Nicholas, 41
Kinmen (Taiwan), 96, 129
kinship, 17, 158, 167
Koch, Robert, 16, 21, 37, 41
Kohn, Eduardo, 39, 171
Korea, 88, 125
Kowloon (Hong Kong), 78, 84, 90, 161
Kourilsky, Philippe, 100–102
Kuo, Steve, 128
kuru, 24

laboratories/laboratory research, 13, 30, 45, 60, 98, 178
Lakoff, Andrew, 44
Lam Chiu Ying, 89, 90, 155, 156, 195n62
Landecker, Hannah, 142
Latham, John, 53, 54
Laver, William Graeme, 42, 43, 49
law, sociology of, 19
Lederberg, Joshua, 38
Leeuwenhoek, Antonie van, 37
Leiris, Michel, 136, 138
Lépinay, Vincent, 114, 115
Le Roy, Charles-Georges, 53
Lever, Ashton, 53, 54
Lévi-Strauss, Claude, 5, 23–28, 60, 61, 134, 170, 174, 176
Lévy-Bruhl, Lucien, 23, 59
Lewis, Daniel, 54, 55
liberal societies, 15
Ligue de Protection des Oiseaux (French League for the Protection of Birds), 53
Limited Test Ban Treaty, 131
Lindenbaum, Shirley, 25
linguistics, 5, 23, 31
Linnaeus, Carl, 53, 54
Lipsitch, Marc, 33
Li Qigui, 74, 75
Liu, Edison, 111
livestock, 81, 83, 139, 140, 156, 158, 165
Lo, Daniel, 162–63
Loeffler, Friedrich, 14
Long Valley (Hong Kong), 89, 154
Los Alamos, 50, 51, 131
lure, 5, 69, 97, 102, 106, 130, 133, 178
Lynteris, Christos, 18

Mackay, George, 140
Mackenzie, Adrian, 112
mad cow disease, 2, 24–28, 174
Mai Po (Hong Kong), 88, 89, 91, 95, 125, 154, 161
Malraux, André, 60
Manchuria, 18
Manhattan project, 131
Mantovani, Alberto, 102
Mao Zedong, 80
markets, 14, 30, 32, 62, 73, 75–80, 82–84, 91, 110, 115, 119, 120, 123, 141, 145, 158, 159, 161–65
Masco, Joseph, 131
Mauss, Marcel, 59
McClintock, Barbara, 99
McClure, Elliott, 89, 92
McCormick, Joseph, 38
McLennan, John Ferguson, 19
melamine crisis, 86
Melville, David, 88
MERS-CoV, 116
Mexico, 30, 64, 111, 141, 146
Migratory Animal Pathological Survey, 89
migratory birds, 2, 43, 88, 91
Millet, Jean, 97
modes of production, 140, 148, 159, 166, 168, 169
Mongolia, 137
monitoring (surveillance), 5, 11, 26, 27, 34, 43, 44, 57, 111, 112, 126, 143, 147, 178
monkeys, 2, 25, 38, 39, 41, 80, 83, 98, 150
Montagnier, Luc, 38
Monte Carlo program, 130
Mooncake game, 129
Moore, Norman, 56
Moss, Stephen, 57

Mount Sinai School of Medicine (New York), 35
Muir, John, 54
Mullis, Kary, 170
Murray Valley Encephalitis virus, 71
Musée de l'Homme, 59–62, 171
Musée des Arts d'Afrique, d'Amérique et d'Océanie, 59
Musée des Arts et Traditions Populaires, 60
Musée d'Ethnographie du Trocadéro, 59
Musée du quai Branly, 59–63
Museum National d'Histoire Naturelle (Paris), 53, 148
museums, 5, 45–49, 54–58, 152, 153, 174
mutations, 3, 5, 11, 16, 29–31, 33, 35, 37, 45, 46, 82, 83, 110, 114, 131, 139, 141–43, 157, 159, 166, 170, 174

Napier, David, 101
National Audubon Society, 53
National Biomedical Research Foundation, 50
National Center for Biotechnology Information, 51
National Institutes of Health (NIH), 35, 40, 51
National Scientific Advisory Board for Biosecurity (NSABB), 33, 35, 40, 158
Natural History Museum of the Smithsonian, 54
nature/naturalism, 14, 15, 19, 22, 23, 26–28, 29, 32, 33, 35, 43, 50–53, 57, 114, 131, 144, 149, 152, 157, 168
Naturschutzbund, 53
Naven ceremony, 134
Newcastle disease, 71, 190n3
Newton, Isaac, 52
Nicholls, John, 85
Nicholson, Edward, 55
Niethammer, Günther, 56
Nipah virus, 41
Nouvel, Jean, 61
nuclear radiation, 130, 175

"One World One Health," 45, 87, 177
ontologies, 28, 39, 46, 103, 123, 143, 164, 166, 168
ornithology, 51–57, 87, 148, 177
Osterhaus, Albert, 40
Osterholm, Michael, 158
Oxford College (Taiwan), 140
Oxford University, 38, 83, 84

Palese, Peter, 35, 36
pandemic influenza (H1N1, H2N2, H3N2), 1, 30, 33, 35–37, 41, 46–49, 81, 87, 111, 120, 143, 157, 158, 171, 172, 178
Papua New Guinea, 24, 134
participation, 11, 12, 15, 17, 18, 177
Pasteur, Louis, 20, 21, 37
Pasteur Institute, 38, 97, 100
Pasteur Research Centre at Hong Kong University, 97
pastoralism, 6, 7, 13, 28, 40, 46, 55, 108, 133, 161, 165, 168, 169, 173, 174, 178
Pedersen, Morten, 40
Peiris, Malik, 83, 85, 86, 90, 101, 105, 106, 193n39, 193n43, 198n96
People for the Ethical Treatment of Animals (PETA), 165
performance, 109, 132, 136
perspective, 3, 4, 7, 12, 20, 26, 27, 36, 38, 39, 55, 61, 65, 81, 99, 108, 109, 135, 155, 174, 175, 178
Peterson, Roger, 152
pets, 123, 133
pharmaceutical industry, 43, 102, 143–46, 171, 172
Philippines, 84, 150
phylogenetic tree, 113
pigeons, 165
pigs, 1, 79, 81–83, 111, 112, 131, 134, 159, 167
plague, 18
play, 109, 115, 123, 125, 133, 136

Plumbbob, Operation, 131
Poon, Leo, 85
Po Toi (Hong Kong), 91
poultry, chickens, ducks, 2, 70–81, 91, 110, 117, 120, 123, 142, 146, 157, 159, 160
precautions, 5, 18, 19, 22, 27, 28, 34, 35, 78, 79, 90, 193n36
preemption, 82, 83, 193n36
preparedness, 1, 3, 5, 7, 23, 45, 49, 57, 58, 59, 62, 82, 86, 128, 138, 165, 173, 177
Preston, Richard, 129
prevention, 4, 5, 7, 14, 23, 27, 45, 46, 49, 57, 62, 165, 173, 177
primitiveness, 14, 17, 22
prions, 25, 26
Prusiner, Stanley, 25
public health, 4, 11, 12, 33, 43, 44, 51, 77, 83, 85, 98, 116, 117, 120, 123, 133, 176

Rabinow, Paul, 170
Ramsar Convention, 88
Ray, John, 53
realism, 116, 117, 130, 132, 133, 135, 138
reciprocity, 168
reserve, 45, 51, 56, 61, 62, 64, 88, 92
reservoir, 1–4, 34, 38, 41, 43, 45, 51, 57, 58, 84, 107, 109, 111, 114, 115, 117, 128, 139, 156, 159, 174
resident birds (endemic species), 52, 92
reverse genetics, 35, 114
Ridgway, Robert, 53, 54
risk, 2, 21, 23, 24, 28, 33–35, 43, 158, 173
rituals, 19, 20, 109, 124, 131, 135, 136, 165
Rivet, Paul, 59
Rivière, Georges-Henri, 59
Robertson Smith, William, 15, 17–19, 21
Rockefeller Institute, 37, 47
Ross River virus, 71
Royal Society for the Protection of Birds, 53

sacred, 15, 16, 19, 20
sacrifice, 12, 16–19, 22, 27, 81, 105, 127, 169, 176, 178, 192n23
Sahlins, Marshall, 168
SARS, 2, 3, 33, 40, 41, 83–87, 90, 98, 102, 109, 111, 115, 116, 122, 128, 174, 193n39, 194n45
Sartre, Jean-Paul, 138, 170
Saudi Arabia, 116
scarcity, 170, 171
scenarios (worst-case scenario, reverse scenario), 83, 108, 109, 114, 117, 118, 120, 128, 131, 132, 136
Schaeffer, Jean-Marie, 130
Schüll, Natasha, 114, 124
Scott, Peter, 56
seagulls, 103, 104
sentinel birds, 69–77, 105–7, 127, 190n4
sentinel cells, 6, 97, 100–103
sentinel posts, 3, 49, 78–87, 147, 174
Severi, Carlo, 134, 135
Severinghaus, Lucia Liu, 93, 94, 152
Severinghaus, Sheldon, 93
shamanism, 39–40, 136
Shau Kei Wan (Hong Kong), 120–21
Shirokogoroff, Sergei, 169
Shortridge, Kennedy, 42, 49, 81, 86, 147
Siberia, 39, 136
signs/signals, 4, 6, 23, 24, 26–28, 31, 34, 35, 39, 49, 69, 97, 98, 100, 101, 103, 106, 122, 123, 125, 133, 164, 174, 175, 178
simulation, 23, 31–35, 41, 49, 62, 108–38, 165, 174, 175
Singapore, 5, 6, 84, 109–19, 153, 174
smallpox, 19, 98, 128
Smith, Derek, 35
Smith, Gavin, 109–14, 132
sociality, 5, 11, 15, 18, 25, 34, 177
Spallanzani, Lazzaro, 37
Spencer, Herbert, 5, 12–16, 21, 23
Spencer, Walter Baldwin, 19
Sri Lanka, 83, 84
stamping out, 24, 27, 143
statistics, 14, 16, 28, 29, 34, 89, 138, 158, 173
Steinman, Ralph, 100
Stépanoff, Charles, 40

St. Jude Hospital (Memphis), 36, 147
stockpiling, 6, 49, 62, 64, 109, 139, 140, 143, 145, 148, 155, 158, 166, 171, 174, 175
storage, 6, 46, 61, 139, 140, 142, 148, 155, 166–67, 171
Strasser, Bruno, 50
Strategic National Stockpile, 143
Sudan, 13
supernatural, belief in the, 17, 22
Swinhoe, Robert, 140, 148

Taiwan, 5, 6, 33, 84, 92, 125, 140–53, 174
Tam Yip Shing, 160
Tan, Desmond, 118
Tan Tock Seng Hospital (Singapore), 115
Taoism, 80, 81, 126, 161
Taubenberger, Jeffery, 35
Tavistock, Marquess of, 55
Testart, Alain, 167–69
Thomas, Keith, 14
To, Johnnie, 164
totemism, 19, 24, 27, 167, 184n49
tracking, 127, 128
Trémon, Anne-Christine, 153
triage, 121, 128, 161
truth, 36, 40, 135, 176
Tsai, Hsiang-Jung, 141–45
toxicity, 103–6, 115
Tumpey, Terrence, 35
Tung Chee-hwa, 86
Tylor, Edward, 17, 27

University Hospital of Bonn, 41
University of California at Davis, 103, 111
University of Hong Kong (HKU), 49, 78, 81, 90, 109, 147, 161
University of Melbourne, 109
University of Shantou, 110, 147

vaccination/vaccines, 5, 11, 13, 20–23, 48, 64, 70, 71, 82, 109, 125, 139, 141–43
value, 6, 106, 138, 148, 154, 165, 171
Vidal, Fernando, 58
Vietnam, 84, 143
virus, 1–5, 14, 37–38, 40, 43, 47, 49, 57, 75, 83, 98, 99, 102, 109, 112, 114, 117, 120, 140, 141, 147, 157, 164, 166, 167, 193nn42–43
veterinary medicine/veterinarians, 13, 24, 41, 43, 81, 83, 159
vom Saal, Frederick, 103

Wain-Hobson, Simon, 34
Wallace, Alfred Russel, 53, 187n65
Walton, Israel, 51, 189n5
Wang, Linfa, 111
Wang, Longde, 119
Wang, Shuheng, 149
Wang, Yichuan, 70, 73, 78, 99
Webster, Robert, 36, 42, 49, 81, 83, 101, 109, 147
Weizmann Institute of Science, 106
Wellcome Trust, 31
Welsh, Geoff, 91–92
Wen Jiabao, 85
West Nile virus, 71, 190n5
White, Gilbert, 51, 189n18
Wild Bird Federation of Taiwan (or Chinese Wild Bird Society), 94, 95, 141, 150
Wild Bird Society of Japan, 150
Wild Bird Society of Taipei, 125, 126
Wild Bird Society of Tainan, 125
Wildfowls and Wetland Trust, 56
Wildlife Conservation Society, 87
Willerslev, Rane, 137
Williams, Greer, 37
Wingspread Declaration, 104
Wolfe, Nathan, 38
Wong, Yoong-Cheong, 116
Woo, John, 164
Woods, Abigail, 13
World Health Organization (WHO), 33, 42, 43, 45–47, 60, 79, 83, 86, 87, 119, 142, 144, 177

World Organization of Animal
Health (OIE), 13, 45, 87, 141–44
World Wildlife Fund, 88
Wu Lien-teh, 18
Wu Hung Chu, 164, 165
Wu Sen-Hsiong, John, 150, 151

Yamashina Institute for Ornithology,
149, 150
Yanni, Carla, 58
Yan Yuren, 75
Yap Him Hoo, 118
yeast, 98
Yersin, Alexandre, 86
Yew, Lee-Kuan, 110, 211n56
You Hanting, 94
Youde, Edward, 88
Yuen Kwok-Yung, 72, 85
Yuen Long (Hong Kong), 70–73

Zahavi, Amotz and Avishag, 104–6
Zhang, Honglin, 85
Zheng Chenggong, 129
Zhong Nanshan, 83
Zigas, Vincent, 24
zoonoses, 2, 18, 181n16
Zylberman, Patrick, 129